THE STEM CELL
CONTROVERSY

THE STEM CELL CONTROVERSY

DEBATING THE ISSUES

EDITED BY

MICHAEL RUSE & CHRISTOPHER A. PYNES

Prometheus Books

59 John Glenn Drive
Amherst, New York 14228-2197

Published 2003 by Prometheus Books

Inquiries should be addressed to
Prometheus Books
59 John Glenn Drive
Amherst, New York 14228–2197
VOICE: 716–691–0133, ext. 207
FAX: 716–564–2711
WWW.PROMETHEUSBOOKS.COM

07 06 05 04 03   5 4 3 2 1

Library of Congress Cataloging-in-Publication Data

The stem cell controversy: debating the issues / edited by Michael Ruse and
    Christopher A. Pynes.
        p. cm.
    Includes bibliographical references.
    ISBN 1–59102–030–1 (alk. paper)
    1. Stem cells—Research—Moral and ethical aspects. I. Ruse, Michael. II.
Pynes, Christopher A., 1970–

QH587.S723 2003
174'.28—dc21

2003041395

Printed in the United States on acid-free paper

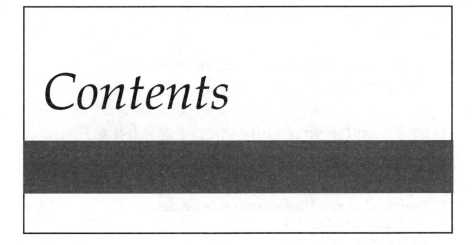

Contents

PART 2. MEDICAL CURES AND PROMISES

PART 3. MORAL ISSUES

PART 5. POLICY ISSUES

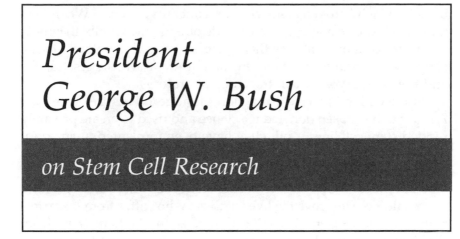

President George W. Bush

on Stem Cell Research

Good evening. I appreciate you giving me a few minutes of your time tonight so I can discuss with you a complex and difficult issue, an issue that is one of the most profound of our time.

The issue of research involving stem cells derived from human embryos is increasingly the subject of a national debate and dinner table discussions. The issue is confronted every day in laboratories as scientists ponder the ethical ramifications of their work. It is agonized over by parents and many couples as they try to have children, or to save children already born.

The issue is debated within the church, with people of different faiths, even many of the same faith coming to different conclusions. Many people are finding that the more they know about stem cell research, the less certain they are about the right ethical and moral conclusions.

My administration must decide whether to allow federal funds, your tax dollars, to be used for scientific research on stem cells derived from human embryos. A large number of these embryos already exist. They are the product of a process called in vitro fertil-

White House news release, August 9, 2001.

ization, which helps so many couples conceive children. When doctors match sperm and egg to create life outside the womb, they usually produce more embryos than are planted in the mother. Once a couple successfully has children, or if they are unsuccessful, the additional embryos remain frozen in laboratories.

Some will not survive during long storage; others are destroyed. A number have been donated to science and used to create privately funded stem cell lines. And a few have been implanted in an adoptive mother and born, and are today healthy children.

Based on preliminary work that has been privately funded, scientists believe further research using stem cells offers great promise that could help improve the lives of those who suffer from many terrible diseases—from juvenile diabetes to Alzheimer's, from Parkinson's to spinal cord injuries. And while scientists admit they are not yet certain, they believe stem cells derived from embryos have unique potential.

You should also know that stem cells can be derived from sources other than embryos—from adult cells, from umbilical cords that are discarded after babies are born, from human placenta. And many scientists feel research on these type of stem cells is also promising. Many patients suffering from a range of diseases are already being helped with treatments developed from adult stem cells.

However, most scientists, at least today, believe that research on embryonic stem cells offer the most promise because these cells have the potential to develop in all of the tissues in the body.

Scientists further believe that rapid progress in this research will come only with federal funds. Federal dollars help attract the best and brightest scientists. They ensure new discoveries are widely shared at the largest number of research facilities and that the research is directed toward the greatest public good.

The United States has a long and proud record of leading the world toward advances in science and medicine that improve human life. And the United States has a long and proud record of upholding the highest standards of ethics as we expand the limits of science and knowledge. Research on embryonic stem cells raises profound ethical questions, because extracting the stem cell destroys the embryo, and thus destroys its potential for life. Like a snowflake, each of these embryos is unique, with the unique genetic potential of an individual human being.

As I thought through this issue, I kept returning to two fundamental questions: First, are these frozen embryos human life, and therefore, something precious to be protected? And second, if they're going to be destroyed anyway, shouldn't they be used for a greater good, for research that has the potential to save and improve other lives?

I've asked those questions and others of scientists, scholars, bioethicists, religious leaders, doctors, researchers, members of Congress, my Cabinet, and my friends. I have read heartfelt letters from many Americans. I have given this issue a great deal of thought, prayer and considerable reflection. And I have found widespread disagreement.

On the first issue, are these embryos human life—well, one researcher told me he believes this five-day-old cluster of cells is not an embryo, not yet an individual, but a pre-embryo. He argued that it has the potential for life, but it is not a life because it cannot develop on its own.

An ethicist dismissed that as a callous attempt at rationalization. Make no mistake, he told me, that cluster of cells is the same way you and I, and all the rest of us, started our lives. One goes with a heavy heart if we use these, he said, because we are dealing with the seeds of the next generation.

And to the other crucial question, if these are going to be destroyed anyway, why not use them for good purpose—I also found different answers. Many argue these embryos are by-products of a process that helps create life, and we should allow couples to donate them to science so they can be used for good purpose instead of wasting their potential. Others will argue there's no such thing as excess life, and the fact that a living being is going to die does not justify experimenting on it or exploiting it as a natural resource.

At its core, this issue forces us to confront fundamental questions about the beginnings of life and the ends of science. It lies at a difficult moral intersection, juxtaposing the need to protect life in all its phases with the prospect of saving and improving life in all its stages.

As the discoveries of modern science create tremendous hope, they also lay vast ethical mine fields. As the genius of science extends the horizons of what we can do, we increasingly confront complex questions about what we should do. We have arrived at that brave new world that seemed so distant in 1932, when Aldous Huxley wrote about human beings created in test tubes in what he called a "hatchery."

In recent weeks, we learned that scientists have created human embryos in test tubes solely to experiment on them. This is deeply troubling, and a warning sign that should prompt all of us to think through these issues very carefully.

Embryonic stem cell research is at the leading edge of a series of moral hazards. The initial stem cell researcher was at first reluctant to begin his research, fearing it might be used for human cloning. Scientists have already cloned a sheep. Researchers are telling us the next step could be to clone human beings to create individual designer stem cells, essentially to grow another you, to be available in case you need another heart or lung or liver.

I strongly oppose human cloning, as do most Americans. We recoil at the idea of growing human beings for spare body parts, or creating life for our convenience. And while we must devote enormous energy to conquering disease, it is equally important that we pay attention to the moral concerns raised by the new frontier of human embryo stem cell research. Even the most noble ends do not justify any means.

My position on these issues is shaped by deeply held beliefs. I'm a strong supporter of science and technology, and believe they have the potential for incredible good—to improve lives, to save life, to conquer disease. Research offers hope that millions of our loved ones may be cured of a disease and rid of their suffering. I have friends whose children suffer from juvenile diabetes. Nancy Reagan has written me about President Reagan's struggle with Alzheimer's. My own family has confronted the tragedy of childhood leukemia. And, like all Americans, I have great hope for cures.

I also believe human life is a sacred gift from our Creator. I worry about a culture that devalues life, and believe as your President I have an important obligation to foster and encourage respect for life in America and throughout the world. And while we're all hopeful about the potential of this research, no one can be certain that the science will live up to the hope it has generated.

Eight years ago, scientists believed fetal tissue research offered great hope for cures and treatments—yet, the progress to date has not lived up to its initial expectations. Embryonic stem cell research offers both great promise and great peril. So I have decided we must proceed with great care.

As a result of private research, more than sixty genetically diverse stem cell lines already exist. They were created from embryos that have already been destroyed, and they have the ability to regenerate themselves indefinitely, creating ongoing opportunities for research. I have concluded that we should allow federal funds to be used for research on these existing stem cell lines, where the life and death decision has already been made.

Leading scientists tell me research on these sixty lines has great promise that could lead to breakthrough therapies and cures. This allows us to explore the promise and potential of stem cell research without crossing a fundamental moral line, by providing taxpayer funding that would sanction or encourage further destruction of human embryos that have at least the potential for life.

I also believe that great scientific progress can be made through aggressive federal funding of research on umbilical cord, placenta, adult, and animal stem cells which do not involve the same moral dilemma. This year, your government will spend $250 million on this important research.

I will also name a president's council to monitor stem cell research, to recommend appropriate guidelines and regulations, and to consider all of the medical and ethical ramifications of biomedical innovation. This council will consist of leading scientists, doctors, ethicists, lawyers, theologians, and others, and will be chaired by Dr. Leon Kass, a leading biomedical ethicist from the University of Chicago.

This council will keep us apprised of new developments and give our nation a forum to continue to discuss and evaluate these important issues. As we go forward, I hope we will always be guided by both intellect and heart, by both our capabilities and our conscience.

I have made this decision with great care, and I pray it is the right one.

Thank you for listening. Good night, and God bless America.

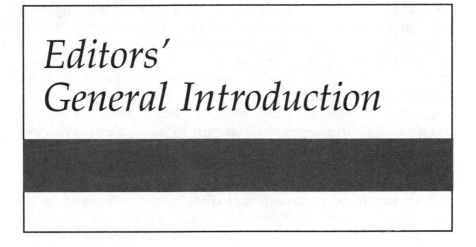

Editors' General Introduction

Parkinson's disease is a truly dreadful ailment that affects 1.5 million Americans. It starts with the death of cells in the brain. This triggers a failure to produce dopamine, a substance needed to transmit information between nerve cells. The end result is the tremors—uncontrollable shaking of the hands—and eventually total rigidity. As the sad story of the actor Michael J. Fox shows only too well, although Parkinson's is mainly an affliction of the aged, it is something that can come down upon the young and is no respecter of status or achievement. There are some ways to slow its course, with drugs that produce dopamine, but these are temporary measures and eventually lose their force. Today, however, there is a whole new avenue of exciting research, an avenue that promises the possibility of complete and lasting relief from this disease. Certain cells of the body, so-called stem cells, have the ability to multiply and to make anew all of the parts of the organism, including the parts of the brain that have ceased to function in those cursed with Parkinson's. With the aim of inducing the brain itself to again produce dopamine, a few medical experts are grafting or injecting stem cells into sufferers. Research is still very much incomplete, but already there are some dramatic successes, with cases of people enjoying at least a 50 percent remission in their symptoms. And these are lasting results, in some cases for ten years or more, with the grafted cells still doing their work.

The way forward is surely obvious. Pour funds into stem cell research. Encourage researchers to turn to its problems. Welcome the interest of the medical profession in such a potentially powerful new way of alleviating disease and misery—for Parkinson's is but one disease that might be cured or moderated by stem cell therapy. Rejoice that human ingenuity and compassion promises to do so much for so many. Unfortunately, things are not this easy nor problem-free. Stem cells do not grow on trees. They are obtained from human beings and—fairly obviously really, when you think about it—although there are various possible sources (like the blood from the umbilical cord of newborns), the most ready font is the very, very young, who have not yet done much developing. It is those cells that still have the greatest potential to transform into the many different parts of the functioning organism. But, of course, the truly very, very young are early embryos, and this at once raises massive ethical and religious problems, not to mention those of social and economic policy.

To put things in context, it can take six or more embryos to produce enough cells to treat a Parkinson's sufferer successfully. Six embryos that have been produced artificially in order to harvest their cells, or six embryos that have been conceived naturally and then aborted, whether or not the abortion was done with the deliberate end of harvesting stem cells. But is it right to treat human beings in this way, or—if you cavil at calling the early embryo a human being—is it right to treat potential human beings in this way? It is this issue that spurred President Clinton in late 1998 to turn for advice to his National Bioethics Advisory Commission (NBAC). It is this issue that President Bush addressed in his talk to the nation on August 9, 2001, as he sought to find a compromise—a middle way—between those for whom stem cell research and therapy is but one more weapon in our fight against ill health and disease, and those for whom any use of embryos, for whatever reason, is absolutely abhorrent and to be forbidden. And it is this issue that we address through the collection of readings in the volume you now hold in your hand. We wanted to give you not the answers, for (even if we could) it is ultimately for you to decide these things for yourself, but at least some of the tools that you might use in order to come to the answers that you find full and satisfying. We start with the science, move then to the potential medical applications, look at both ethical and religious issues, and conclude with some writings about policy and the possible ways forward.

This is an explosive issue that is going to get more so. The deliberate destruction of embryos—abortion—has split families, societies, nations. Thus far, one of the most important elements in the controversy has been the fact that, with obvious exceptions, the act of abortion usually occurs in a haze of moral disapproval. Or, at least, it occurs under circumstances and for reasons in which those who want to find moral objections can succeed in their quest. Unwanted pregnancies do not occur by chance but through carelessness or the like. Why then should a human or potential human be destroyed just to mop up after someone who has failed to exercise restraint or caution? But the stem cell issue brings in another dimension entirely. If individuals fall sick, it is usually not their fault. Why then should they be denied the opportunity of cure? Why then should they be denied the opportunity of cure if already there is an abundant supply of the needed material anyway—the millions of aborted fetuses that are already discarded every year? By what right should those who suffer be denied cure by the wishes or demands of those whose moral or religious convictions are not shared by the rest of us?

On these issues we have striven deliberately, not to be neutral in the sense of uncaring, but to be balanced in the sense of letting people of very different beliefs and persuasions all have their say. Even the science itself is not without controversy, for (as you will see) there are debates about the possible sources of stem cells. Potential applications raise even more questions and disagreements, hopes and false dreams. And then, as we move to ethical or moral issues, conflict really escalates, as people draw diametrically opposed conclusions about right courses of action. Expectedly, religion does little to calm these conflicts, and between adherents of different faiths there is at times stark opposition. Which means that, at the end, policymakers have to pick up the pieces and to make recommendations, to enact policies, and to try to enforce and to follow them. Hence the kind of proposal put forward by President Bush—to allow some stem cell research from already-existing lines, but not to allow (certainly not to fund by the government) the deliberate production of new sources of such cells.

We have chosen readings that are accessible to the nonspecialist, and to further aid comprehension we have added a glossary of unfamiliar terms and concepts. Some of the ideas in the section on ethics may be a little unfamiliar—unfamiliar, but not really that complicated—especially if you have never taken a course in philosophy. Just remember that people who deal with moral issues generally try to relate their thinking to a background theory or world picture. Thus, for instance, the Christian dealing with issues of world poverty would try to understand the issues in terms of Jesus' preaching in the Sermon on the Mount, as well as through the parables (like that of the Good Samaritan), together with the writings of great Christian thinkers from Saint Paul right down to our own day. Those who take a more secular approach to morality—that is, those who try to understand right and wrong without direct reference to a deity (although they may in fact themselves be deeply religious)—tend to fall into one of two camps. Either they are "deontologists"—judging matters by intention—or they are "teleologists" (or, to use another name for the latter, "consequentialists")—judging matters by results. Of course, often intentions and results come to the same thing—a person who sets out to do something is more likely to achieve the end than one who has no such aims—but intentions and results can come apart, and then real differences can emerge. Particularly the deontologist will stress the overall attitude that one has to people and their status and needs, whereas the consequentialist will stress the end result, downplaying the ways used to achieve that end.

The greatest of the deontologists was the late-eighteenth–century German philosopher Immanuel Kant, who argued that acts should be judged by what he called the "Categorical Imperative." The easiest formulation of this rule is that one should always treat people for their own sake, as ends in themselves, and never as means to some other good or end. Unless the victim in some way benefits as well, making an example of someone just for the general good is never justifiable. The greatest of the consequentialists were the British utilitarians, notably in the nineteenth century John Stuart Mill, who argued that one judges right actions by the results, especially the extent to which such actions maximize happiness and minimize unhappiness. One has to balance the benefits of one person or group against the benefits of other persons or groups. You might think that a Kantian would automatically come out against stem cell research, for surely in such an enterprise you are treating the embryos as a means to the ends of others. You would not balance the medical benefits of Nazi human experimentations against the suffering of the innocents on whom Dr. Mengele and other medical ghouls operated. Conversely, you might think that a utilitarian would automatically come out in favor of stem cell research, for the happiness of the sufferer will be eased and relieved and this will outweigh the ill or harm to the unaware embryo. Most of us dislike the vivisection of dumb animals, but we swallow our scruples and let it happen. But, as you will learn, although there is undoubted truth in this expectation—Kantians do tend to be more conservative on moral issues, and utilitarians do tend to be more liberal about these things—matters are not entirely simple. Is the fetus really a human being, and if not, then does the Categorical Imperative apply to it? Conversely, is the happiness of the relieved sufferer enough to outweigh the potential happiness of the fetus? And in any case, what about the happiness of those people concerned for and involved with the embryo?

These are some of the questions you will see asked; and as you will see, different people come to different conclusions. Your conclusions must be your conclusions. What we hope is that this collection will help you to make them, for decisions must be made. As the great French philosopher Jean-Paul Sartre once said, not to chose a course of action is in fact to chose another course of action. If you do not think about these issues, then implicitly you are letting others do the thinking for you, and their conclusions are your conclusions, and their actions are your actions. And with such an important matter as stem cell research, this should not be so. As the great Anglican divine and poet John Donne once said: "No man is an island." We are all part of society and what happens to one of us happens to all of us. So join with us in reading what we have collected, and in thinking and reasoning about these very important issues—important for us and for our fellow citizens and indeed for all members of the human species.

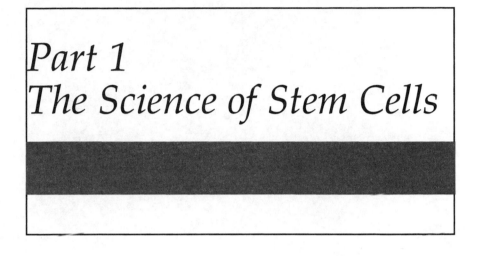

Part 1
The Science of Stem Cells

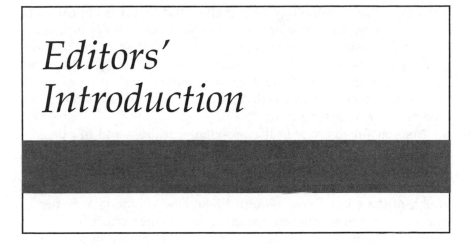

Editors' Introduction

To begin at the beginning, we need to know something about the science. All living organisms are made of cells, some being but one cell and others (including the higher animals) of many different cells, each doing its allotted task or function. In sexual organisms like humans, the male and the female each contribute a special sex cell (the sperm and the ovum, respectively) that fuse together (into the zygote) that then divides and multiplies as the individual grows and develops. The early cells are known as stem cells, meaning that they can go on dividing indefinitely and that they have the potential to make all of the specialized cells needed in the body. We start our collection with a primer put out by the National Institutes of Health in May of 2000 that gives you the basics of the science, showing how stem cells arise and what their fates and potentialities hold. This will give you the basic science and much of the needed technical terminology, although you should then supplement this with the next reading, a specially commissioned piece by historian and philosopher of science Jane Maienschein, who digs into the past, showing that concepts and understanding only evolve gradually, that terminology does matter, and that it is important to know precisely what you are arguing about before you plunge into heated discussion. Sometimes people object that scholars, philosophers particularly, are overly concerned with language when they should be getting on with the facts, and there is perhaps some truth in this complaint. But words do matter, and a lot of heat later can be avoided by a little light now.

As you already saw in the president's address and in our general introduction, the really hard issues—moral and religious—come from the use of fetuses as a source of stem cells. Obviously if one could avoid or minimize the use of early embryonic material much of the controversy would die and be avoided. A matter of some considerable interest therefore is whether one can harvest stem cells from older organisms—that is, from grown humans who would not be destroyed in the process of stem cell recovery and who could give their informed consent to the donation. At the moment, however, it is a hotly contested issue as to whether it is possible to use adults as a source for all of the stem cells that seem useful and needed. To give you some understanding of the science, we present two readings, one somewhat pessimistic and another (specially commissioned from Sidney Houff of the Loyola University Medical Center in Chicago) that thinks the prospects are rather more promising. Then finally we present a fascinating news report by *Washington Post* science reporter Rick Weiss. He discusses the possibility of stimulating cells to reproduce themselves and obtaining stem cells in that fashion. We see at once how important are some of Maienschein's concerns about language and definitions. These artificially stimulated cells—"parthenots"—are not like normal embryos, and some would argue that therefore they have no rights and raise no special moral and religious questions. Others would argue precisely the opposite. So much is going to depend on precisely what you mean by "normal," "potential," "embryo," "human," and other related terms.

1
Stem Cells

A Primer

National Institutes of Health

T his primer presents background information on stem cells. It includes an explanation of what stem cells are; what pluripotent stem cells are; how pluripotent stem cells are derived; why pluripotent stem cells are important to science; why they hold such great promise for advances in health care; and what adult stem cells are.

Recent published reports on the isolation and successful culturing of the first human pluripotent stem cell lines have generated great excitement and have brought biomedical research to the edge of a new frontier. The development of these human pluripotent stem cell lines deserves close scientific examination, evaluation of the promise for new therapies, and prevention strategies, and open discussion of the ethical issues.

In order to understand the importance of this discovery as well as the related scientific, medical, and ethical issues, it is absolutely essential to first clarify the terms and definitions.

From the National Institutes of Health, May 2000.

DEFINITIONS

DNA—abbreviation for deoxyribonucleic acid which makes up genes.

Gene—a functional unit of heredity which is a segment of DNA located in a specific site on a chromosome. A gene directs the formation of an enzyme or other protein.

Somatic cell—cell of the body other than egg or sperm.

Somatic cell nuclear transfer—the transfer of a cell nucleus from a somatic cell into an egg from which the nucleus has been removed.

Stem cells—cells that have the ability to divide for indefinite periods in culture and to give rise to specialized cells.

Pluripotent—capable of giving rise to most tissues of an organism.

Totipotent—having unlimited capability. Totipotent cells have the capacity to specialize into extraembryonic membranes and tissues, the embryo, and all postembryonic tissues and organs.

WHAT IS A STEM CELL?

Stem cells have the ability to divide for indefinite periods in culture and to give rise to specialized cells. They are best described in the context of normal human development. Human development begins when a sperm fertilizes an egg and creates a single cell that has the potential to form an entire organism. This fertilized egg is *totipotent*, meaning that its potential is total. In the first hours after fertilization, this cell divides into identical totipotent cells. This means that either one of these cells, if placed into a woman's uterus, has the potential to develop into a fetus. In fact, identical twins develop when two totipotent cells separate and develop into two individual, genetically identical human beings. Approximately four days after fertilization and after several cycles of cell division, these totipotent cells begin to specialize, forming a hollow sphere of cells, called a blastocyst. The blastocyst has an outer layer of cells and inside the hollow sphere, there is a cluster of cells called the inner cell mass.

The outer layer of cells will go on to form the placenta and other supporting tissues needed for fetal development in the uterus. The inner cell mass cells will go on to form virtually all of the tissues of the human body. Although the inner cell mass cells can form virtually every type of cell found in the human body, they cannot form an organism because they are unable to give rise to the placenta and supporting tissues necessary for development in the human uterus. These inner cell mass cells are *pluripotent*—they can give rise to many types of cells but not all types of cells necessary for fetal development. Because their potential is not total, they are not totipotent and they are not embryos. In fact, if an inner cell mass cell were placed into a woman's uterus, it would not develop into a fetus.

The pluripotent stem cells undergo further specialization into stem cells that are committed to give rise to cells that have a particular function. Examples of this include blood stem cells which give rise to red blood cells, white blood cells, and platelets; and skin stem cells that give rise to the various types of skin cells. These more specialized stem cells are called *multipotent*.

While stem cells are extraordinarily important in early human development, multipotent stem cells are also found in children and adults. For example, consider one of the best understood stem cells, the blood stem cell. Blood stem cells reside in the bone marrow of every child and adult, and in fact, they can be found in very small numbers circulating in the blood stream. Blood stem cells perform the critical role of continually replenishing our supply of blood cells—red blood cells, white blood cells, and platelets—throughout life. A person cannot survive without blood stem cells.

HOW ARE PLURIPOTENT STEM CELLS DERIVED?

At present, human pluripotent cell lines have been developed from two sources[1] with methods previously developed in work with animal models.

(1) In the work done by Dr. [James] Thomson, pluripotent stem cells were isolated directly from the inner cell mass of human embryos at the blastocyst stage. Dr. Thomson received embryos from IVF (In Vitro Fertilization) clinics—these embryos were in excess of the clinical need for infertility treatment. The embryos were made for purposes of reproduction, not research. Informed consent was obtained from the donor couples. Dr. Thomson isolated

the inner cell mass and cultured these cells producing a pluripotent stem cell line.

(2) In contrast, Dr. [John] Gearhart isolated pluripotent stem cells from fetal tissue obtained from terminated pregnancies. Informed consent was obtained from the donors after they had independently made the decision to terminate their pregnancy. Dr. Gearhart took cells from the region of the fetus that was destined to develop into the testes or the ovaries. Although the cells developed in Dr. Gearhart's lab and Dr. Thomson's lab were derived from different sources, they appear to be very similar.

The use of somatic cell nuclear transfer (SCNT) may be another way that pluripotent stem cells could be isolated. In studies with animals using SCNT, researchers take a normal animal egg cell and remove the nucleus (cell structure containing the chromosomes). The material left behind in the egg cell contains nutrients and other energy-producing materials that are essential for embryo development. Then, using carefully worked out laboratory conditions, a somatic cell—any cell other than an egg or a sperm cell—is placed next to the egg from which the nucleus had been removed, and the two are fused. The resulting fused cell, and its immediate descendants, are believed to have the full potential to develop into an entire animal, and hence are totipotent. As described these totipotent cells will soon form a blastocyst. Cells from the inner cell mass of this blastocyst could, in theory, be used to develop pluripotent stem cell lines. Indeed, any method by which a human blastocyst is formed could potentially serve as a source of human pluripotent stem cells.

POTENTIAL APPLICATIONS OF PLURIPOTENT STEM CELLS

There are several important reasons why the isolation of human pluripotent stem cells is important to science and to advances in health care. At the most fundamental level, pluripotent stem cells could help us to understand the complex events that occur during human development. A primary goal of this work would be the identification of the factors involved in the cellular decision-making process that results in cell specialization. We know that turning genes on and off is central to this process, but we do not know much about these "decision-making" genes or what turns them on or off. Some of our most serious medical conditions, such as cancer and birth defects, are due to abnormal cell specialization and cell division. A better understanding of normal cell processes will allow us to further delineate the fundamental errors that cause these often deadly illnesses.

Human pluripotent stem cell research could also dramatically change the way we develop drugs and test them for safety. For example, new medications could be initially tested using human cell lines. Cell lines are currently used in this way (for example, cancer cells). Pluripotent stem cells would allow testing in more cell types. This would not replace testing in whole animals and testing in human beings, but it would streamline the process of drug development. Only the drugs that are both safe and appear to have a beneficial effect in cell line testing would graduate to further testing in laboratory animals and human subjects.

Perhaps the most far-reaching potential application of human pluripotent stem cells is the generation of cells and tissue that could be used for so-called cell therapies. Many diseases and disorders result from disruption of cellular function or destruction of tissues of the body. Today, donated organs and tissues are often used to replace ailing or destroyed tissue. Unfortunately, the number of people suffering from these disorders far outstrips the number of organs available for transplantation. Pluripotent stem cells, stimulated to develop into specialized cells, offer the possibility of a renewable source of replacement cells and tissue to treat a myriad of diseases, conditions, and disabilities including Parkinson's and Alzheimer's diseases, spinal cord injury, stroke, burns, heart disease, diabetes, osteoarthritis, and rheumatoid arthritis. There is almost no realm of medicine that might not be touched by this innovation. Some details of two of these examples follow.

- Transplant of healthy heart muscle cells could provide new hope for patients with chronic heart disease whose hearts can no longer pump adequately. The hope is to develop heart muscle cells from human pluripotent stem cells and transplant them into the failing heart muscle in order to augment the function of the failing heart. Preliminary work in mice and other animals has demonstrated that healthy heart muscle cells transplanted into the heart successfully repopulate the heart tissue and work together with the host cells. These experiments show that this type of transplantation is feasible.

- In the many individuals who suffer from Type I diabetes, the production of insulin by specialized pancreatic cells, called islet cells, is disrupted. There is evidence that transplantation of either the entire pancreas or isolated islet cells could mitigate the need for insulin injections. Islet cell lines derived from human pluripotent stem cells could be used for diabetes research and, ultimately, for transplantation.

While this research shows extraordinary promise, there is much to be done before we can realize these innovations. Technological challenges remain before these discoveries can be incorporated into clinical practice. These challenges, though significant, are not insurmountable.

First, we must do the basic research to understand the cellular events that lead to cell specialization in the human, so that we can direct these pluripotent stem cells to become the type(s) of tissue needed for transplantation.

Second, before we can use these cells for transplantation, we must overcome the well-known problem of immune rejection. Because human pluripotent stem cells derived from embryos or fetal tissue would be genetically different from the recipient, future research would need to focus on modifying human pluripotent stem cells to minimize tissue incompatibility or to create tissue banks with the most common tissue-type profiles.

The use of somatic cell nuclear transfer (SCNT) would be another way to overcome the problem of tissue incompatibility for some patients. For example, consider a person with progressive heart failure. Using SCNT, the nucleus of virtually any somatic cell from that patient could be fused with a donor egg cell from which the nucleus had been removed. With proper stimulation the cell would develop into a blastocyst: cells from the inner cell mass could be taken to create a culture of pluripotent cells. These cells could then be stimulated to develop into heart muscle cells. Because the vast majority of genetic information is contained in the nucleus, these cells would be essentially identical genetically to the person with the failing heart. When these heart muscle cells were transplanted back into the patient, there would likely be no rejection and no need to expose the patient to immune-suppressing drugs, which can have toxic effects.

ADULT STEM CELLS

As noted earlier, multipotent stem cells can be found in some types of adult tissue. In fact, stem cells are needed to replenish the supply cells in our body that normally wear out. An example, which was mentioned previously, is the blood stem cell.

Multipotent stem cells have not been found for all types of adult tissue, but discoveries in this area of research are increasing. For example, until recently, it was thought that stem cells were not present in the adult nervous system, but, in recent years, neuronal stem cells have been isolated from the rat and mouse nervous systems. The experience in humans is more limited. In humans, neuronal stem cells have been isolated from fetal tissue and a kind of cell that may be a neuronal stem cell has been isolated from adult brain tissue that was surgically removed for the treatment of epilepsy.

Do Adult Stem Cells Have the Same Potential as Pluripotent Stem Cells?

Until recently, there was little evidence in mammals that multipotent cells such as blood stem cells could change course and produce skin cells, liver cells or any cell other than a blood stem cell or a specific type of blood cell; however, research in animals is leading scientists to question this view.

In animals, it has been shown that some adult stem cells previously thought to be committed to the development of one line of specialized cells are able to develop into other types of specialized cells. For example, recent experiments in mice suggest that when neural stem cells were placed into the bone marrow, they appeared to produce a variety of blood cell types. In addition, studies with rats have indicated that stem cells found in the bone marrow were able to produce liver cells. These exciting findings suggest that even after a stem cell has begun to specialize, the stem cell may, under certain conditions, be more flexible than first thought. At this time, demonstration of the flexibility of adult stem cells has been only observed in animals and limited to a few tissue types.

Why Not Just Pursue Research with Adult Stem Cells?

Research on human adult stem cells suggests that these multipotent cells have great potential for use in both research and in the development of cell therapies. For example, there would be many advantages to using adult stem cells for transplantation. If we could isolate the adult stem cells from a patient, coax them to divide and direct their specialization and then transplant them back into the patient, it is unlikely that such cells would be rejected. The use of adult stem cells for such cell therapies would certainly reduce or even avoid the practice of using stem cells that were derived from human embryos or human fetal tissue, sources that trouble many people on ethical grounds.

While adult stem cells hold real promise, there are some significant limitations to what we may or may not be able to accomplish with them. First of all, stem cells from adults have not been isolated for all tissues of the body. Although many different kinds of multipotent stem cells have been identified, adult stem cells for all cell and tissue types have not yet been found in the adult human. For example, we have not located adult cardiac stem cells or adult pancreatic islet stem cells in humans.

Second, adult stem cells are often present in only minute quantities, are difficult to isolate and purify, and their numbers may decrease with age. For example, brain cells from adults that may be neuronal stem cells have only been obtained by removing a portion of the brain of epileptics, not a trivial procedure.

Any attempt to use stem cells from a patient's own body for treatment would require that stem cells would first have to be isolated from the patient and then grown in culture in sufficient numbers to obtain adequate quantities for treatment. For some acute disorders, there may not be enough time to grow enough cells to use for treatment. In other disorders, caused by a genetic defect, the genetic error would likely be present in the patient's stem cells. Cells from such a patient may not be appropriate for transplantation. There is evidence that stem cells from adults may have not have the same capacity to proliferate as younger cells do. In addition, adult stem cells may contain more DNA abnormalities, caused by exposure to daily living, including sunlight, toxins, and by expected errors made in DNA replication during the course of a lifetime. These potential weaknesses could limit the usefulness of adult stem cells.

Research on the early stages of cell specialization may not be possible with adult stem cells since they appear to be farther along the specialization pathway than pluripotent stem cells. In addition, one adult stem cell line may be able to form several, perhaps 3 or 4, tissue types, but there is no clear evidence that stem cells from adults, human or animal, are pluripotent. In fact, there is no evidence that adult stem cells have the broad potential characteristic of pluripotent stem cells. In order to determine the very best source of many of the specialized cells and tissues of the body for new treatments and even cures, it will be vitally important to study the developmental potential of adult stem cells and compare it to that of pluripotent stem cells.

SUMMARY

Given the enormous promise of stem cells to the development of new therapies for the most devastating diseases, it is important to simultaneously pursue all lines of research. Science and scientists need to search for the very best sources of these cells. When they are identified, regardless of their sources, researchers will use them to pursue the development of new cell therapies.

The development of stem cell lines, both pluripotent and multipotent, that may produce many tissues of the human body is an important scientific breakthrough. It is not too unrealistic to say that this research has the potential to revolutionize the practice of medicine and improve the quality and length of life.

NOTE

1. Michael Shamblott et al., "Derivation of Pluripotent Stem Cells from Cultured Human Primordial Germ Cells," *Proceedings of the National Academy of Sciences USA* 95 (1998): 13726–31; James Thomson et al., "Embryonic Stem Cell Lines Derived from Human Blastocysts," *Science* 282 (1998): 1145–47.

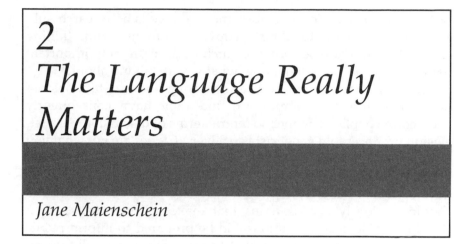

2
The Language Really Matters

Jane Maienschein

O n July 31, 2001, the U.S. House of Representatives passed the "Human Cloning Prohibition Act." On August 9, President Bush announced his decision to allow limited research on existing stem cell lines but not on "embryos." In contrast, the U.K. has explicitly authorized "therapeutic cloning." In the blizzard of media attention, much has been said about bioethical, legal, and social implications of these decisions and about whether and how to modify them. The Senate can decline to act on cloning, Congress can decide to overrule the President's Executive order concerning federal funding, and the U.K. can take advantage of U.S. hesitations to ensure their lead in these areas of research. In the meantime, privately funded research and the work that goes on every day in fertility clinics can continue untouched by federal policy.

Throughout these discussions, and despite the NIH's excellent attempts to provide information and education, subtleties of the science have received much less careful consideration.[1] Legislators and reporters struggle with fundamental terms such as "cloning," "pluripotency," "stem cells," and "embryos," as well as whether

"adult" are preferable to "embryonic" stem cells as research subjects. They profess to abhor "copying humans" and "killing
embryos" but to see scientific research as an extremely important
path to progress. Do we all know what we are talking about?

Scientists have largely resisted worrying about definitions not
deemed scientifically important, and they have consequently
allowed the public to define the central terms of discussion. As a tel-
evision producer told geneticist Lee Silver, "Clones are not what you
think they are." Silver concluded that "Science and scientists would
be better served by choosing other words to explain advances in
developmental biotechnology to the public."[2] I disagree. Scientists
should obviously choose terms that make scientific sense, then
should provide clear definitions and be prepared to inform public
debate. Without informed scientists' perspectives, we risk making
bad policies, in some cases without even realizing that we are doing
so because we thought we were talking about something else. In
addition, definitions change. They are a matter of convention among
a relevant community, and can change either because what we know
changes or because the community changes. This is the case for
embryos, cloning, and stem cells—for each of which it matters very
much what we mean, and for each of which the meanings have
changed over times and in different communities. What follows
here is a discussion of these terms, in historical context.

EMBRYOS

The first *Encyclopedia Britannica* (1771) defined an "embrio" as "the
first rudiments of an animal in the womb, before the several members are distinctly formed; after which period it is denominated a
foetus." The word is said to have come from a combination of bryein
and em, for swelling inside. This fit a favorite late-eighteenth-cen-
tury theory that organisms begin essentially as preformed encapsu-
lated forms that swell and grow larger. It was also compatible with
the alternative theory of epigenesis, or gradual emergence of form,
typically thought to be guided by some vital force or directive
agency in the Aristotelian sense of a becoming that actualizes pre-
existing potential. Both theories regarded the embryo as an incipient
organism, requiring time and nutrition to become an adult. It was

also generally assumed that at some point in the process of development "quickening" occurs and indicates the beginning of life. The quickening seemed clearly to be the "real" starting point for individual development.

Several steps informed scientific understanding of embryos. In 1827, Karl Ernst von Baer discovered that mammals begin as eggs, as birds so manifestly do, and hence that mammals have continuity with previous generations through that material egg. Despite biologists' best efforts, they could see little about the development of those eggs hidden away inside the mammalian mother until the mid-nineteenth century brought advances in microscopy and histological methods for fixing, staining, and microtoming. During the 1870s, Wilhelm His developed a system for serially sectioning embryos that allowed study of internal structure and organization and led to his pioneering set of human developmental stages.[3] While biologists used the term "embryo" for all stages following fertilization, they saw the stages after formation of the germ layers as the important ones for development of individual organisms, and the earliest stages as largely a matter of cell division and nutrition. The fact that typically the dividing egg remains about the same size up through the blastocyst stage, despite cell divisions, reinforces this interpretation.

By the early twentieth century, biologists had modified both preformationism (now seen in terms of predetermined inherited material passed on from parents to offspring through inherited particles called genes) and epigenesis (the material unfolding of preorganized "organ-forming germ-regions" or the expression of relevant "fates" by cells and germ layers). Determining the balance of predirection and epigenetic emergence was, as Oskar Hertwig put it, *The Biological Problem of To-day.*[4] Biologists prodded, sliced, and diced developing organisms from the germ cell stage through fertilization, morula, blastula, gastrula, and subsequent stages to uncover the processes and patterns of morphogenesis. Along the way, it became much less important what they called an "embryo" or what the domain of "embryology" was thought to be. Usually, the terms referred loosely to all development from fertilization on, but the important stages were thought to be post-gastrulation, which brings the formation of three germ layers.

Whereas leading morphologists including His, Ernst Haeckel, and others emphasized the significance of those germ layers as the starting point for differentiation and individuation, others began to look at a diversity of animals to push back discussions to fertilization and the very first cell division as also important for morphogenesis. Cell lineage, for example, involved tracing lineages of each cell from the unfertilized egg cell itself, though each division until the multiplication of cells made it impossible to follow the details any longer. With meticulous methods for preserving each stage (the egg itself, two cells, four cells, etc.), then observing and drawing or photographing the details, researchers discovered that from the beginning some species have very determinate cleavage patterns suggesting an early differentiation. In contrast, other species allow considerable regulation and demonstrate variability in developmental patterns.

Clearly, morphogenesis is a complex process, with variations across species and even within species depending on the details of an individual case. Throughout this period, morphogenesis, or genesis of form, remained the primary focus of research. The "embryo" receded into the background in importance as a sort of placeholder, while the series of defined stages and the patterns and processes of change gained importance.

One episode still appears in textbooks and class discussions, though we can learn new lessons from the case. Wilhelm Roux and Hans Driesch each experimented on the two-cell stage of an organism: Roux on frogs and Driesch on sea urchins. Roux killed one of the two cells with a hot needle and discovered that the remaining cell develops essentially as it would if it were part of the normal embryo. He concluded that development occurs in a mosaic manner, with individual cells adopting individual fates in the whole organism that follows. Driesch took the two-celled stage of his sea urchins and shook apart the two blastomeres, watching each grow into a larval form as if it were a reorganized whole and not just one half of a mosaic. Driesch concluded that each blastomere at the two and even four cell stage is "totipotent" and capable of considerable regulation in response to changing conditions surrounding the cell.[5]

This difference in their results raised fundamental questions about the nature of differentiation, and led other biologists to take up parallel studies in other organisms for comparison. Some are more mosaic-like, and others more regulative, they discovered. In the period between 1880 and 1910, researchers made tremendous advances in understanding cells and development.[6] Yet they soon reached the limits of those lines of research, and had to set aside what they recognized were exciting questions about regulation and the limits of such totipotency and regulation.

Ross Harrison pursued new methods of exploration by culturing tissues outside the body for the first time; he cultured frog neuroblasts in a lymph medium, and thereby carried out a first step toward current research on precursor and stem cells. Harrison and Hans Spemann also transplanted various parts of donor organisms onto other hosts to discover the relative contributions of each, and this led Spemann to his concept of the "organizer" at the gastrula stage.[7] During this time, deciding what was meant by an "embryo" or "organism" became less important than determining to what extent the stages were defined, predictable, and determined *vs.* regulated in response to environmental conditions.

Around the same time, the ninth *Britannica* (1897) defined "embryology" as "somewhat vaguely applied to the product of generation of any plant or animal which is in process of formation." The term included all stages from fertilized eggs to birth, though a human embryo was seen as becoming a "fetus" at 8 weeks, when it "has assumed the characteristic form and structure of the parent." Textbooks of human embryology outlined successive stages, focusing on appearance of organs and visible features such as limbs and face. Both because of what was possible with existing research approaches and because of what medicine deemed important, later fetal stages remained the primary focus.

We see this for humans in Keibel and Mall's edited classic 1910 *Manual of Human Embryology*. The authors admitted that they knew little of the early stages: "Nothing is known concerning the fertilization of the human ovum, but it may be presumed that it takes place in essentially the same manner as in other animals." "The segmentation stages of the human ovum have not yet been observed." And so on. They were restricted to what they could see with the few selected specimens they had, and these represented later stages, starting with the fourteen day "ovum," which itself revealed little. In later stages they could focus on recognizable organs and the sequence in which they arise. Yet even here, the authors noted,

It is clear that the normal fully developed organism cannot be produced without a certain regular succession in the development of the organs and a regular interdependence of the individual devel-

opmental processes, but it is open to question whether this interdependence is the result of the individual and independent development of each organ taking place in such a way that it fits into
that of the others or whether it is due to the individual anlagen of
an organism mutually influencing one another during the development so as to cause the formation of a normal organism.[8]

These leading human embryologists used the terms "ovum" and
"organism," while focusing on the succession of parts and organ.
They largely yielded the term "embryo" to non-scientific usage. Biologists instead considered "morphogenesis," "organogenesis,"
"growth," "differentiation," and such processes. By the 1960s,
"embryology" had largely disappeared along with discussion of
embryos, giving way to "developmental biology." Encyclopedia definitions also became looser, so that the *Britannica*'s current "embryo"
is "the early developmental stage of an animal while it is in the egg
or within the uterus of the mother. In humans the term is applied to
the unborn child until the end of the seventh week following conception; from the eighth week the unborn child is called a fetus."

This looseness of terminology began to matter a great deal more
for political, legal, and social reasons after 1978, when Patrick
Steptoe's team demonstrated the astonishing efficacy of *in vitro* fertilization. They could fertilize human eggs in the laboratory, implant
them, and watch mothers bring them to term. They learned to nurture the eggs in glass dishes, select the most viable, freeze and save
them, and to manipulate these eggs in many ways. Because of the
evident therapeutic benefits to otherwise infertile patients, and perhaps because Steptoe did not announce his efforts until they had
been successful, the public accepted this innovation as a medical
advance.

The Vatican was not happy, seeing such manipulation of even
early developmental stages as morally illicit since Pope Pius IX had
in 1869 declared that "ensoulment" and hence life begins at conception. As Karen Dawson noted rather understatedly in a 1990 edited
volume on *Embryo Experimentation*, "The debate about embryo
research is far from finished and will continue to occur whether governments seek to regulate this rapidly developing area of science."[9]
Nonetheless, public opinion in the western world has accepted *in
vitro* medical therapies to allow more people to bear children. Jewish
and Muslim scholars, for example, regard life as starting later (forty

days is the most commonly accepted time). Typically, the term "embryo" has applied to the stages from fertilization through the eighth week in humans, but there is clearly an important qualitative biological difference between the blastula and gastrula stages.

The U.S. Congress has passed legislation to protect human embryos, the most recent laws prohibiting use of federal funds for "the creation of a human embryo for research purposes or for research in which a human embryo is destroyed." In the Fiscal Year 1998 Appropriations Act, they explicitly defined human embryos to "include any organism, not protected as a human subject under 45 CFR 46 as of the date of the enactment of this Act, that is derived by fertilization, parthenogenesis, cloning, or any other means from one or more human gametes or human diploid cells." Fertilization was clearly thought to be key. Yet it is difficult to discover any clearly reasoned argument leading to this decision, nor any biologically informed discussion of whether subtle distinctions ought to be allowed. Instead, wrapping this decision into omnibus appropriations acts seems to have allowed it to slide through without significant debate. There is room for more careful consideration of what we mean, why, and with what implications.

Up through the blastocyst stage, it is possible to culture the human egg in a glass dish, outside the body. After gastrulation, at least with current technology, it is not. Up until that stage, the blastocyst cells remain largely "plastic" and at least multi or pluripotent (capable of becoming more than one or all cell types). In mice, for example, it has proven possible to combine pairs or small clusters of cells from different organisms and have them join and develop together normally, as a sort of hybrid or chimeric new organism. By gastrulation, this regulative response is no longer possible, at least with present technologies and understanding. The cells have by then begun to differentiate and to become channeled into their respective fate paths.

There is a difference between the blastula and gastrula stages, then, and it is important. Yet we have not developed conventionally accepted terms for these differences. The NIH refers to the entire process as the embryo, while President Bush followed many others in referring to the preimplantation blastula as a "pre-embryo." Clearly, it was important to him to make the distinction. The pre-embryo is different from an embryo: perhaps it is not "alive" in the same way, perhaps it is not an "individual," or conceptually not yet a "potential human" in the same way that an implanted and differentiated postgastrula embryo is. As R. J. Blandon put it in 1971, summarizing Reichert's 1873 studies, the blastocyst is that "structure existing during the period between the end of cleavage and the attachment of the embryo to the uterine wall." It is not itself "the embryo, he declaims, but only the developmental stage of a new the creature, which is the embryo."[10]

Clearly there are biological differences, and much hinges politically on our definitions. Defining a pregastrulation embryo as something different may be the most useful approach. This is certainly the line taken by the British Parliament in 1990 when they agreed to allow some research on the grounds that "such embryos have not progressed to the point that they can be considered as individuals."[11] We could follow Lee Silver's suggestion and seek different terms altogether, but those will not inform the public discussion that relies on familiar and conventionally accepted language. Therefore, we need to work harder to clarify our definitions, to recognize that these will change over time as we learn more, and to develop clearer and scientifically defensible connections between the biological and public language usage. Historians, philosophers, and bioethicists can help with that process.

CLONING

When Ian Wilmut and the Roslin Institute team announced that they had cloned a sheep using a new technique, people around the world scrambled to understand what somatic cell nuclear transfer really means.[12] The image of babies as miniature copies of parents, or as time-delayed twins disturbed many people. After all, the child is not supposed to be the brother of the man—or woman. The collective, if sometimes fascinated, "yuk" suggested that people were not ready for this. And so, as many governments worked to determine whether and what regulations should be imposed, and as Ian Wilmut himself took a lead in speaking out against cloning humans, the U.S. Congress began to generate its own proposed legislation.

Congressman Vernon Ehlers, himself a Ph.D. physicist and one of the few Congressmen with any significant scientific training, proposed a very clean and simple bill that would have outlawed the use of federal funds to clone humans for the purposes of creating a human being. Numerous other bills also appeared, in various stages of development, and a few made their way to committees. I was working as science advisor to our Congressman at the time, and was astonished that some of the early bills were so vaguely worded that they would have outlawed gene cloning or even, in the most extreme case, much of agriculture or even identical twins since one early draft bill proposed to prohibit any genetic replication. Fortunately, in the course of a series of hearings and staff briefings, it became clear that the Congressmen did not know what they were talking about sufficiently to take action. When the Food and Drug Administration declared that they would exercise control over any possible cloning and therefore that no special law was needed, there was a sigh of relief. This held off the perceived need to rush to action. It also helped that Wilmut kept reminding people how difficult the cloning process was, and how many failures occurred before achieving that initial success.

What became very clear very quickly, nonetheless, was the importance of words and of perspective. The concept of cloning was not new. "Cloning" actually first appeared in agriculture, when Herbert John Webber wrote in 1903 of "clons" that were "groups of plants that are propagated by the use of any form of vegetative parts such as bulbs, tubers, cuttings, grafts etc. and are simply parts of the same individual."[13] By the 1920s, cloning covered a variety of asexually reproducing genetic copies, including tree runners, regenerating worms, bacterial divisions, and cell copying in cultures.

In the 1950s, Robert Briggs and Thomas J. King successfully transplanted nuclei from donor *Xenopus laevis* into host frogs. This was cloning through nuclear transplantation, but they could not achieve success with nuclei taken from post-gastrula stages.[14] In the 1960s, John Gurdon significantly expanded the range of techniques and cells, including differentiated frog somatic cells, that could be cloned this way. The work was technically challenging, yielding a low success rate, and attempts to transfer the techniques to mammals failed. The successes showed that there were no natural barriers to such hybrids, but the prospects for human cloning seemed very remote. A brief media blitz gave way to skepticism, while the public probably assumed that such cloning would not happen and relatively few scientists continued to pursue the possibilities.

Textbooks suggested that there was no evidence that it was impossible to do such cloning in mammals, and even humans. James Watson, John Tooze, and David Jurtz wrote in their *Recombinant DNA: A Short Course* that "In the immediate future there is little likelihood of nuclear transplantation being attempted with any other mammalian species" or even much beyond frogs. "If the efficiency and reproducibility can be improved, the method may, however, find a place in animal breeding. In theory it could be attempted with human eggs and embryonic cells, but for what reason? There is no practical application."[15] That was 1983, and only two years after the first American human *in vitro* fertilization, so the enormous possibilities in terms of enhancing human fertility were not yet apparent.

The 1950s and 1960s brought cloning of cell lines, with the recognition that there is tremendous advantage in having a large collection of genetically identical cells: both to control experimental variables and to have sufficient material should therapeutic uses be developed. This cell culturing research built on the earlier tissue culture studies that had begun decades earlier in developmental biology. The 1960s brought gene cloning, with the same advantages of genetic duplication. Progressively faster, more accurate, and more efficient laboratory techniques and microbial "factories" have made gene cloning a familiar activity for undergraduate scientists today. The advances have carried us far beyond what Briggs and King could have imagined, but the concepts remain the same. Cloning consistently refers to the full range of types of genetic copying, whatever way that is carried out and on whatever organisms.

Notwithstanding the many efforts to explain that genetic replication does not mean that the expressed offspring cells, tissues, or organisms will be exact copies of the original, the popular perception still sees it that way. As Silver noted, cloning escaped the biologists because of Dolly [the sheep] and the subsequent furor and took on its own public meaning. This does not mean that biologists should give cloning up to the popular press, however. Rather scientists need to continue informing discussions and helping public figures make the distinctions necessary to ground sound policy and legal rulings.

The House bill H.R. 2505, "Human Cloning Prohibition Act of 2001," defines human cloning as "human asexual reproduction, accomplished by introducing nuclear material from one or more human somatic cells into a fertilized or unfertilized oocyte whose nuclear material has been removed or inactivated so as to produce a living organism (at any stage of development) that is genetically virtually identical to an existing or previously existing human organism." Somatic cell "means a diploid cell (having a complete set of chromosomes) obtained or derived from a living or deceased human body at any stage of development." With those definitions, the proposed law states that "It shall be unlawful for any person or entity, public or private, in or affecting interstate commerce, knowingly (1) to perform or attempt to perform human cloning; (2) to participate in an attempt to perform human cloning; or (3) to ship or receive for any purpose an embryo produced by human cloning or any produce derived from such embryo."[16] This bill was passed by the House in part to set the boundaries for decisions about stem cell research and to make it clear that even if such research were allowed, human cloning for the purposes of reproducing humans would be prohibited. To date, the Senate has not enacted any similar bill and seems unlikely to do so without further provocation.

What is important is the boundary constraints that Congress is trying to establish. Far from that initial frightened reaction to the idea of runaway Frankenstein-like scientists copying people in their labs, a more reflective consideration has led to an idea of cloning in terms of somatic cell nuclear transfer. Either the Congress is not aware of the possibility or does not find worrisome that a researcher might transfer a nucleus (or even entire cell) from, for example, a blastocyst into another oocyte from which the nucleus had been removed. This would involve nuclear transfer, but not technically somatic cell nuclear transfer. It could be cloning in the biological sense, since the goal would be genetic copying by the resultant cells. But it is not cloning in the sense of H.R. 2505. Perhaps the Congressmen and their staffs meant to cover this contingency with the definition that somatic cells are derived from the human body during any stage of development. But technically, the blastocyst is not yet a "body," so that still leaves ambiguity.

Cloning clearly includes a range of activities, and that range is likely to expand as we learn more. It is important not to rush to judgment to ban all possible currently foreseeable activities, since we will surely predict imperfectly, nor to allow everything just because we cannot develop crisp definitions. It is also important that scientists not give up the discussions because they are difficult and not central to carrying out the work of each individual scientist. In the long run, it does matter. If legislators are going to make decisions that will restrict some scientific research, then scientists had best be involved in informing those decisions. Historians, philosophers, and bioethicists can help provide perspective, demonstrate what is and is not new, what is and is not good scientific research, and what ethical, legal, and social implications are likely to result.

STEM CELLS

As with embryos and cloning, the term "stem cells" has taken on public meaning. Yet here, perhaps since there is no immediate intuitive sense of the term, that public discussion had been more closely informed by scientific explanation. The primary association is with embryos, since "embryonic stem cells" have moved into the forefront of public debate quite recently.

Also as with embryos and cloning, stem cells and research on them are not new. Yet as with cloning, technical innovations in recent years have excited public imagination. With cloning, Dolly evoked images of sheep-like genetically copied humans. With embryonic stem cell research, the images are of killed embryos on the one hand or of magical laboratory-generated miracle cures on the other. Neither is realistic, nor even close to the whole story.

When Driesch discovered that each of two, or even four separated blastomeres could develop into a whole sea urchin larva, he demonstrated that each cell retains a totipotency, or ability to make the whole. As early twentieth century research placed the material of inheritance in genes, arranged in chromosomes, and passed along through cell and nuclear division to each daughter cell, this raised what Thomas Hunt Morgan called a paradox of development. If all the cells within an organism have the same genetic constitution, what causes them to differentiate? And to what extent is the differentiation determined, or flexible in response to different environmental conditions? This set of questions, essentially a new form of the preformation-epigenesis debates has dominated developmental biology. Exploration of this set of issues lies at the root of stem cell research.

Some cells are able to give rise to an entire, normal, healthy organism. These are totipotent cells, and the fertilized egg is the most obvious. Under some conditions, each of up to the eight cells in humans can do the same, yielding identical octuplets. Again focusing on humans, until the blastocyst stage, there is no reason to believe that any cells are fully differentiated into different cell types. Indeed, the form "blasto" as in blastomere, blastocyst, or blastoderm was used widely by the late nineteenth century for stems, sprouts, or germs, in the sense of able to give rise to form. During the blastocyst stage, it begins to be the case that each cell can no longer give rise to the whole organism by itself. Instead of totipotency, therefore, some of these cells are pluripotent or multipotent, and in later developmental stages following gastrulation, they become differentiated as unipotent, or precursor cells for some particular cell type. Distinctions in these terms are important, and much political ferment hinges on them.

Pluripotent or stem cells are those capable of becoming many, some would say all, of the different cell types of the normal body. This does not mean, however, that any one stem cell can become all the types. Indeed, it cannot. There is no evidence of the capacity and no actual cases in which stem cells give rise to entire organisms. Stem cells, therefore, are not tiny embryos in the rough, nor can they become embryos. They are not "alive" in the sense of giving rise to entire living organisms, though they are capable of reproducing themselves. True pluripotent stem cells can replicate, and can give rise to lines of future stem cells as well as to differentiated cells of a variety of types. Current definitions hold pluripotent stem cells as able to give rise to "most" or "every" cell type, and at least to some types of cells from each of the three germ layers. In practice, however, it is very difficult to prove that any line of stem cells can, in fact, actually give rise to *every* type of cell. Theoretically, however, that is believed possible.

In the 1950s, F. C. Stewart showed that he could generate a whole carrot from single root cells, thereby demonstrating totipotency and raising once again the question just how flexible the most "plastic" cells are. Nuclear transplantation demonstrated a great deal of adaptability in frogs. And, following the discovery in the 1960s that bone marrow transplants allowed patients to produce healthy blood cells to treat anemia and leukemia, the 1960s brought major advances in understanding the multipotent haemotopoietic stem cells. These cells carry out two functions: they differentiate into normal blood cells, and they also maintain and renew the population of stem cells. J. P. Lewis and F. R. Trobaugh argued that they therefore satisfied the definition of stem cells.[17] Their cells were self-renewing as well as differentiating into blood cells, and hence gave rise to two cell types. They were therefore multipotent, and considered true stem cells. From the 1960s until recently, such multipotent bone marrow cell lines have provided our best known example of stem cells. They later proved also to differentiate into other cells useful to the immune system, and therefore to have considerable therapeutic potential as well as demonstrating wider multipotency than originally thought.

Discussions of such stem cells formed a basis for discussion of regeneration. How is it that some organisms, some of the time, and for only some cells, are capable of regenerating injured parts? Crabs can regenerate a claw, but humans cannot regenerate a hand. Worms can regenerate most of their segmented bodies, and snakes regularly regenerate their skins. Why, how, when, under what conditions, and can we use an understanding of such processes to develop medically useful applications, researchers asked. Biologists, who had previously assumed that differentiation works in only one direction as daughter cells cannot go back to a more general state, occasionally questioned that assumption. As Elizabeth Hays wrote in 1974,

> Until better experimental evidence is forthcoming, it no longer is useful, however, to think of differentiation as an irreversible condition, even though we know that somatic cell differentiation is very stable. . . . This is difficult to do because our concepts so overlap that we almost invariably think of differentiation as a restriction of developmental potency, even though we know that cells like the ovum that are truly totipotent actually are differentiated, highly specialized cells.[18]

This was a period of trying to sort out what was meant by stem cells, precursor cells, differentiated and undifferentiated cells, and the extent to which the normal condition could be overruled by experimental manipulation. It was the time when the groundwork for our current discussions began. Unfortunately, it was also a time when some of the definitions remained murky. Whether a stem cell must be multipotent or pluripotent, and what difference that makes were such core questions. What has changed in recent decades is, first, the discovery that stem cells seem to have far wider capabilities than previously thought and, second, that embryonic stem cells appear to have the greatest capacities and there are vast collections of embryos in storage in fertility clinics.

In 1994, the NIH Human Embryo Research Panel recognized the potential in human embryo research, but nonetheless called for moral caution in pursuing this study. The debates might have stalled there, but for the independent publications by James Thomson and John Gearhart in 1998.[19] They showed that stem cell lines could be cultured and could be caused to differentiate into a variety of cell types. As reporter Gretchen Vogel summarized in a special stem cell issue of *Science*, "Conventional wisdom has assumed that once a cell has been programmed to produce a particular tissue, its fate was sealed, and it could not reprogram itself to make another tissue. But in the last year, a number of studies have surprised scientists by showing that stem cells from one tissue, such as brain, could change into another, such as blood." These discoveries served to "signal the unexpected power of stem cells."[20] As leading stem cell researchers hasten to point out, even as they argue for federal funding of their research, many open questions remain: including how much stem cells can be made to do, and whether the most controversial stem cells derived from embryos will prove sufficiently more efficacious to overrule the moral concerns. That we will never learn if we do not ask remains very clear. And that we will not be certain what we have learned unless we are very careful to define our terms and to develop a shared accumulation of knowledge is equally clear.

CONCLUSIONS

Suddenly in the past few years, cloning and stem cell discoveries have heated political discussions about when and what to regulate, and where to draw lines on what procedures scientists should be allowed to pursue. As seen in the preceding discussion, it very much matters what the underlying terms mean. If the public decides to protect embryos, we had better know whether we mean by that fertilized eggs, blastocysts, gastrulas, or only post-implantation stages. Bush and his advisors refer to "pre-embryos," "frozen embryos" stored in fertility clinics, and researchable cell lines from embryos already "destroyed." Bush's August 9 televised speech suggested that he sees cell lines as derived from "destroyed embryos," as already non-living, and therefore as legitimate research subjects, whereas frozen fertilized eggs have a different status, as live embryos. Others would disagree, but in diverse ways. We see considerable reliance on fuzzy definitions to make biologically questionable distinctions that then have tremendous political force.

Clearly it matters what we count as an embryo, what it means to destroy one, and what capacities each developmental stage holds. It matters what we mean by cloning, and how we distinguish embryonic and adult stem cells, and whether they must be pluripotent or multipotent to have therapeutic value. If world leaders and legislators place restrictions on the pursuit of scientific knowledge, the scientific community must work hard to insure the best possible understanding of the science involved. Historians, philosophers, and bioethicists must also work hard and must avoid being seduced by facile distinctions and false hopes growing out of uncorrected misunderstandings and misinterpretations. It is vitally important to define terms clearly and carefully, and for at least a representative set of experts to engage effectively in public discussion not just of the details of their own science but of the words used and the implications of the choices made.

NOTES

1. The National Institutes of Health report, "Stem Cells: Scientific Progress and Future Research Directions," was released in July 2001 in response to a request from Secretary of Health and Human Services Tommy Thompson. The 11 August 2001, *New York Times* includes a photo with a caption, "Karen P. Hughes, counselor to the president, yesterday holding a book that the president used in making his stem cell decision." The NIH Web site contains very valuable and reliable "primers" on stem cell research and other difficult health-related topics.

2. L. Silver, "What Are Clones: They're Not What You Think They Are," *Nature* 412 (July 5, 2001): 21.

3. W. His, *Anatomie Menschlicher Embryonen. I. Embryonen des Ersten Monats* (Leipzig, Ger.: FCW Vogel, 1880).

4. O. Hertwig, *The Biological Problem of Today, Preformation or Epigenesis? A Basis of the Organic Theory*, trans. P. C. Mitchell (London: William

Heinemann, 1896). First published as O. Hertwig, *Zeit- und Streitfragen der Biologie* (Jena, Ger.: Fischer, 1894).

5. J. Maienschein, "The Origins of Entwicklungsmechanik," in *A Conceptual History of Modern Embryology*, ed. Scott Gilbert (Baltimore: Johns Hopkins University Press, 1994), pp. 43–61.

6. E. B. Wilson, *The Cell in Development and Inheritance*, 2d ed. 1900, 3d ed. 1925 (New York: Macmillan, 1896).

7. V. Hamburger, *The Heritage of Experimental Embryology* (New York: Oxford University Press, 1988).

8. F. Keibel and F. P. Mall, eds., *Manual of Human Embryology* (Philadelphia and London: Lippincott, 1910), pp. xvi, 18, 980; R. O'Rahilly and F. Müller, *Developmental Stages in Human Embryos* (Washington, D.C.: Carnegie Institution of Washington Publications, 1987), p. 637, includes a revision of Streeter's "Horizons" and a survey of the Carnegie Collection.

9. P. Singer et al., eds., *Embryo Experimentation* (Cambridge: Cambridge University Press, 1990), p. xvi.

10. R. J. Blandon, *The Biology of the Blastocyst* (Chicago: University of Chicago Press, 1971).

11. "The Meaning of Life," *Nature* 412 (19 July 2001): 255.

12. I. Wilmut et al., "Viable Offspring Derived from Fetal and Adult Mammalian Cells," *Nature* 385 (1997): 810–13.

13. H. J. Webber, "New Horticultural and Agricultural Terms," *Science* 18 (1903): 501–503.

14. R. Briggs and T. J. King, "Transplantation of Living Nuclei from Blastula Cells into Enucleated Frogs' Eggs," *Proceedings of the National Academy of Sciences USA* 38 (1952): 455–63.

15. J. D. Watson, J. Tooze, and D. T. Kurtz, *Recombinant DNA: A Short Course* (New York: Scientific American Books, 1983), pp. 207–208.

16. Congressional Record [online], http://thomas.loc.gov [1 August 2001].

17. J. P. Lewis and F. E. Trobaugh, "Haematopoietic Stem Cells," *Nature* 204 (1964): 589–90.

18. E. Hays, "Cellular Basis of Regeneration," in *Concepts of Development*, ed. J. Lash and J. R. Wittaker (Sunderland, Mass.: Sinauer Associates, 1974), p. 405.

19. J. A. Thomson et al., "Embryonic Stem Cell Lines Derived from Human Blastocysts," *Science* 282 (1998): 1145–47; J. Gearhart, "New Potential for Human Embryonic Stem Cells," *Science* 282 (1998): 1061–62.

20. G. Vogel, "Can Old Cells Learn New Tricks?" *Science* 287 (2000): 1418.

3
Can Old Cells Learn New Tricks?

Stems cells found in adults show surprising versatility. But it's not clear whether they can match the power of cells from embryos.

Gretchen Vogel

Stem cell biologist Margaret Goodell has never seen her work on muscle and blood development as particularly political, so she was surprised when last month the Coalition of Americans for Research Ethics (CARE), a group that opposes the use of embryos in research, invited her to speak at a congressional briefing in Washington, D.C. She was even more astonished to find herself quoted by conservative columnist George Will a few weeks later.

Goodell gained this sudden notoriety because her work, and that of other teams around the world, just might provide a way around the moral and political quagmire that has engulfed stem cell research to date. Since their discovery in 1998, human embryonic stem cells have been one of the hottest scientific properties around. Because these cells can theoretically be coaxed to differentiate into any type of cell in the body, they open up tantalizing possibilities, such as lab-grown tissues or even replacement organs to treat a variety of human ills, from diabetes to Alzheimer's. Polit-ically, however, human stem cells have been a much tougher sell, as they are derived from embryos or fetuses. Indeed, most research is on hold as policymakers grapple with the ethics of human embryo research.

Enter Goodell, whose work suggests that stem cells derived from adults, in this case, from mouse muscle biopsies, can perform many of the same tricks as embryonic stem (ES) cells can—but without the ethical baggage. Both CARE and George Will seized upon her work as an indication that research on ES cells could remain on hold with no appreciable loss to medicine. "There's a lot less moral ambiguity about the adult stem cells," says bioethicist and CARE member Kevin Fitzgerald of Loyola University Medical Center in Chicago.

But can adult stem cells really fulfill the same potential as embryonic stem cells can? At this stage, the answer is by no means clear. Indeed, scientists caution that it is too early to know if even ES cells will produce the cornucopia of new tissues and organs that some envision. "It is still early days in the human embryonic stem cell world," says stem cell biologist Daniel Marshak of Osiris Ther-apeutics in Baltimore, which works with adult-derived stem cells.

From a scientific standpoint, adult and embryonic stem cells both have distinct benefits and drawbacks. And harnessing either one will be tough. Although scientists have been working with mouse ES cells for two decades, most work has focused on creating transgenic mice rather than creating lab-grown tissues. Only a handful of groups around the world have discovered how to nudge the cells toward certain desired fates. But that work gained new prominence in late 1998, when two independent teams, led by James Thomson of the University of Wisconsin, Madison, and John Gearhart of The Johns Hopkins University, announced they could grow human stem cells in culture. Suddenly the work in mouse cells could be applied to human cells—in the hope of curing disease.

The beauty of embryonic stem cells lies in their malleability. One of their defining characteristics is their ability to differentiate into any cell type. Indeed, researchers have shown that they can get mouse ES cells to differentiate in lab culture into various tissues, including brain cells and pancreatic cells.

Studies with rodents also indicated that cells derived from ES cells could restore certain missing nerve functions, suggesting the possibility of treating neurological disorders. Last summer, Oliver Brüstle of the University of Bonn Medical Center and Ronald McKay of the U.S. National Institute of Neurological Disorders and Stroke and their colleagues reported that they could coax mouse ES cells to become glial cells, a type of neuronal support cell that produces the neuron-protecting myelin sheath. When the team then injected these cells into the brains of mice that lacked myelin, the transplants produced normal-looking myelin.[1] And in December, a team led by Dennis Choi and John McDonald at Washington University School of Medicine in St. Louis showed that immature nerve cells that were generated from mouse ES cells and transplanted into the damaged spinal cords of rats partially restored the animals' spinal cord function.[2] Although no one has yet published evidence that human ES cells can achieve similar feats, Gearhart says he is working with several groups at Johns Hopkins to test the abilities of his cells in animal models of spinal cord injury and neurodegenerative diseases, including amyotrophic lateral sclerosis and Parkinson's disease.

While Gearhart and his colleagues were grappling with ES cells, Goodell and others were concentrating on adult stem cells. Conventional wisdom had assumed that once a cell had been programmed to produce a particular tissue, its fate was sealed, and it could not reprogram itself to make another tissue. But in the last year, a number of studies have surprised scientists by showing that stem cells from one tissue, such as brain, could change into another, such as blood.[3] Evidence is mounting that the findings are not aberrations but may signal the unexpected power of adult stem cells. For example, Goodell and her colleagues, prompted by the discovery of blood-forming brain cells, found that cells from mouse muscle could repopulate the bloodstream and rescue mice that had received an otherwise lethal dose of radiation.

Bone marrow stem cells may be even more versatile. At the American Society of Hematology meeting in December, hematologist Catherine Verfaillie of the University of Minnesota, Minneapolis, reported that she has isolated cells from the bone marrow of children and adults that seem to have an amazing range of abilities. For instance, Verfaillie and graduate student Morayma Reyes have evidence that the cells can become brain cells and liver cell precursors, plus all three kinds of muscle—heart, skeletal, and smooth. "They are almost like ES cells," she says, in their ability to form different cell types.

These malleable bone marrow cells are rare, Verfaillie admits. She estimates that perhaps one in 10 billion marrow cells has such versatility. And they are only recognizable by their abilities; the team has not yet found a molecular marker that distinguishes the unusually powerful cells from other bone marrow cells. Still, she says, her team has isolated "a handful" of such cells from 80 percent of the bone marrow samples they've taken. Although the versatile cells are more plentiful in children, Verfaillie's team has also found them in donors between forty-five and fifty years old.

Verfaillie's work has not yet been published nor her observations replicated. Even so, many researchers are excited by the work. The cells "look extremely interesting," says hematologist and stem cell researcher Leonard Zon of Children's Hospital in Boston. Stem cell biologist Ihor Lemischka of Princeton University agrees. "I'm very intrigued," he says, although he cautions that data from one lab should not outweigh the decades of research on mouse ES cells.

Besides skirting the ethical dilemmas surrounding research on embryonic and fetal stem cells, adult cells like Verfaillie's might have another advantage: They may be easier to manage. ES cells tend to differentiate spontaneously into all kinds of tissue. When injected under the skin of immune-compromised mice, for example, they grow into teratomas—tumors consisting of numerous cell types, from gut to skin. Before applying the cells in human disease, researchers will have to learn how to get them to produce only the desired cell types. "You don't want teeth or bone in your brain. You don't want muscle in your liver," says stem cell researcher Evan Snyder of Children's Hospital in Boston. In contrast, Verfaillie says her cells are "better behaved." They do not spontaneously differentiate but can be induced to do so by applying appropriate growth factors or other external cues.

Adult stem cells have a drawback, however, in that some seem to lose their ability to divide and differentiate after a time in culture. This short lifespan might make them unsuitable for some medical applications. By contrast, mouse ES cells have a long track record in the lab, says Goodell, and so far it seems that they "are truly infinite in their capacity to divide. There are [mouse] cell lines that have been around for ten years, and there is no evidence that they have lost their 'stem cell-ness' or their potency," she says.

For these and other reasons, many researchers say, adult-derived stem cells are not going to be an exact substitute for embryonic or fetal cells. "There are adult cell types that may have the potential to repopulate a number of different types of tissues," says Goodell. "But that does not mean they are ES cells. Embryonic stem cells have great potential. The last thing we should do is restrict research." Right now, she says, stem cell specialists want to study both adult and embryonic stem cells to find out just what their capabilities might be.

That may be difficult. At the moment, human ES cells are unavailable to most researchers because of proprietary concerns and the uncertain legal status of the cells. Internationally, most research on human ES cells is on hold while legislatures and funding agencies wrestle with the ethical issues. In the United States, the National Institutes of Health is the government agency that would fund the research, and currently, researchers are not allowed to use NIH funds for work with human ES cells. Many European countries, too, are still developing new policies on the use of the cells.[4]

The final version of NIH's guidelines for use of embryonic and fetal stem cells will not appear before early summer, says Lana Skirboll, NIH associate director for science policy. The draft guidelines would allow use of NIH funds for ES cell research as long as the derivation of the cells, by private institutions, met certain ethical standards.[5] But several members of Congress are considering legislation that would overrule the guidelines and block federal funding of ES cell research. At least some of that debate is likely to focus on whether adult stem cells do in fact have the potential to do as much as their embryonic precursors.

NOTES

1. Oliver Brüstle et al., "Embryonic Stem Cell-Derived Glial Precursors: A Source of Myelinating Transplants," *Science* 285 (1999): 754.

2. Ingrid Wickelgren, "Rat Spinal Cord Function Partially Restored," *Science* 286 (1999): 1826.

3. Christopher R. R. Bjornson et al., "Turning Brain into Blood: A Hematopoietic Fate Adopted by Adult Neural Stem Cells in Vivo," *Science* 283 (1999): 534.

4. Noëlle Lenoir, "Viewpoint," *Science* 287 (2000): 1425.

5. Gretchen Vogel, "NIH Sets Rules for Funding Embryonic Stem Cell Research," *Science* 286 (1999): 2050.

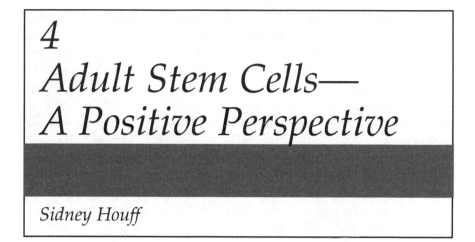

4
Adult Stem Cells— A Positive Perspective

Sidney Houff

The use of stem cells to replace adult cells injured by trauma or disease has generated considerable interest. Given their undifferentiated state and proliferate capacity, embryonic stem cells have been the focus of most of the research in cell transplantation. The use of embryonic stem cells has resulted in ethical and legal controversies. These issues resulted in President Bush's decision to limit federal funding to research only existing embryonic stem cells.

Cell transplantation using adult stem cells would circumvent the legal and ethical issues entailed in the use of fetal tissues. But can adult stem cells do the job? I think they can, at least in part. In addition, the knowledge we gain from studying adult stem cells could have a significant impact on embryonic stem cell research.

Do adult stem cells exist and, if so, do they have the same properties found in embryonic stem cells? Adult stem cells have been isolated from a number of organs including bone marrow and brains of rodents, nonhuman primates, and humans. In rodent experiments, differentiation of adult stem cells has resulted in adult cells of the brain as well as other organs. Adult hemapoetic cells have been reported to arise from differentiation of adult stem cells isolated from brain. In humans, donor-derived hepatocytes have been found in the liver of recipients of bone marrow transplants. Further studies have shown that recipients of liver transplants can generate hepatocytes which populated the transplanted liver. These studies suggest the adult animals and humans harbor pluripotential cells capable of generating adult cells that are incorporated into both simple and highly complex cellular architectures. The results of these and other experiments using adult stem cells need confirmation. As with many of the studies with embryonic stem cells, the functional capability of cells derived from adult stem cells needs study.

How do adult stem cells compare to embryonic stem cells? Operationally, stem cells have been defined as being able to differentiate into a wide variety of adult cell types while generating additional stem cells. In many studies, adult stem cells appear not appear to have the ability to generate as many adult cell types or have the proliferative capacity of embryonic stem cells. These findings have suggested that adult stem cells are likely not to be as useful as embryonic stem cells in cell transplantation paradigms. I will propose here that these properties of adult stem cells may actually benefit from our approach to cell transplantation.

Stem cell differentiation into adult cell types occurs through a series of cell divisions associated with increasing phenotypic specificity. If adult stem cells have undergone cell division and differentiation, this could limit the possible adult cell phenotypes they are capable of generating. These partially differentiated stem cells have been called progenitor cells to separate them from true stem cells.

Progenitor cells could offer several advantages in cell transplantation. There are fewer steps to generate the adult cell phenotype needed for tissue repair. Progenitor cells are likely to be more responsive to molecular signals present in adult tissues. The prior commitment of progenitor cells may also facilitate the integration of derived adult cells into the functional architecture of the adult tissue. These properties are likely to be of special importance in transplantation of the central nervous system where anatomical location and cell phenotype are critical for successful neural function. Adult stem cells may offer advantages here as well.

The stem cell "niche" has been emphasized in many studies of the biology of stem cell development. The cellular environment is critically important in stem cell biology. Although the "niche" has yet to be completly defined, molecular signals such as growth factors, cell adhesion molecules, and other molecular signals are known to be important in initiating and supporting stem cell development into adult cell types.

Stem cells, or progenitor cells, present in adult tissues are likely to be more competent to respond to the signals present in the adult stem cell "niche" when compared to embryonic stem cells. It may even be possible that embryonic stem cells or their early derivatives may not have the ability to respond to signals present in adult tissues or require additional signals that are no longer present in adult tissues. Here, the adult stem cell or progenitor is likely to be more successful in repairing tissue damage in the adult. Furthermore, a better understanding of the molecular signals present in the adult stem cell "niche" may be important in understanding cell differentiation beginning with embryonic stem cells.

The study of adult stem cells is likely to lead to many answers important in fulfilling the promise of stem cell biology. They may be able to fulfill the role of tissue repair themselves. The use of adult stem cells, besides avoiding the ethical and legal problems associated with the use of embryonic cells, offer clear advantages in cell transplantation paradigms. The future of stem cell biology, if it is to reach its full potential, will require the study of adult stem cells along with the more conventional approaches using embryonic stem cells.

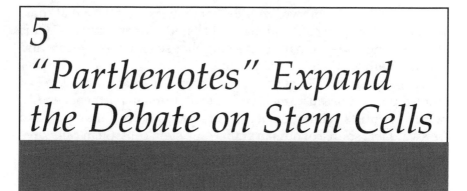

5
"Parthenotes" Expand the Debate on Stem Cells

Rick Weiss

Whenis an embryo not an embryo? It's a question fit for a Zen master, but it's one that will have to be answered by federal bureaucrats, researchers, and ethicists as biologists explore the weird world of parthenogenesis.

Parthenogenesis, from the Greek word for "virgin birth," is an unusual mode of procreation in which, to put it plainly, the male might as well find something else to do. Female aphids and turkeys, and certain female reptiles, are among the critters that can reproduce this way. Their eggs can divide on their own as though they had been fertilized by a sperm, then go on to develop into embryos and offspring.

Using chemicals that mimic a sperm's arrival, scientists in recent years have triggered parthenogenesis in the eggs of a few mammals, including rabbits and mice, but the resulting "embryo" has never developed beyond the early fetus stage. Now, with the announcement two weeks ago of the first successful induction of parthenogenesis in human eggs, the Zen riddle arises: Is the resulting doomed entity an embryo, with the same moral standing that some people confer upon conventional and cloned human embryos? Or is it something else—perhaps something that scientists can create and destroy with legal and moral impunity?

This question calls out for an answer—at least on the level of public policy—because the products of human parthenogenesis, which scientists call not embryos but "parthenotes," may prove to be a valuable source of human embryonic stem cells. Stem cells have the capacity to morph into all kinds of cells and tissues, and may someday prove useful in the treatment of degenerative diseases such as diabetes and Parkinson's. There is a roaring political and ethical debate about the use of human embryos as sources of stem cells. Should a similar debate surround work on parthenotes, although there is overwhelming evidence that they lack any potential to become people?

"It forces everybody to get far more nuanced in their thinking about what it is about early life forms that, in their minds, renders them morally significant," said R. Alta Charo, a professor of law and bioethics at the University of Wisconsin in Madison.

The new work was conducted by Michael West and colleagues at Advanced Cell Technology Inc. in Worcester, Massachusetts. In one report, published in an online journal on November 25, the team said it had used chemicals to stimulate human eggs to grow into embryo-like balls of about one hundred cells—the first creation of human parthenotes. (The stimulus was applied before the egg underwent the normal ejection of half its chromosomes, which typically occurs at the time of fertilization to accommodate the sperm's DNA.)

Inexplicably, those parthenotes did not contain any stem cells. But related research, described by West at a scientific meeting on December 2, suggests the goal of obtaining stem cells from human parthenotes is achievable. The team made parthenotes from monkey eggs—the first such success in our near-human relatives—and retrieved from one of those parthenotes what appeared to be stem cells.

Those cells turned into intestine, skeletal muscle, retina, hair follicles, cartilage, bone, and other cell types—even heart cells beating in unison. Some turned into nerve cells that secreted the brain chemical dopamine, the kind of cell that is gradually lost by Parkinson's patients. The idea, West said, is to grow replacement cells and tissues from a patient's own eggs so they are genetically so similar to the woman that they won't be rejected by her immune system.

Men might also someday benefit from parthenote-based therapies, scientists said. Research in animals suggests that male parthenotes can be made by inserting two sperm into an egg whose own DNA has been removed, then stimulating that reconstructed egg to start dividing. Stem cells from the resulting parthenote would be near genetic matches to the male patient who had contributed the sperm.

Technical problems still loom large. No one knows, for example, whether parthenote cells are dangerously inbred because all their genes come from only one parent. "Are they completely normal? Are they prone to metastasize like tumor cells? There's really not enough information," said George Seidel, an embryologist at Colorado State University in Fort Collins. In mammals, both male and female chromosomes are required for normal development.

A practical complication, said Johns Hopkins stem cell researcher John Gearhart, is that viable eggs are difficult to retrieve from postmenopausal women—who are more likely than younger women to be suffering from degenerative diseases.

But Jerry Hall, laboratory director at the Institute for Reproductive Medicine and Genetic Testing in Westwood, California, said his research on mouse parthenotes suggests the cells have promise. "We need to prove they are safe and effective," Hall said. "But I can tell you, they look very normal."

Even if the science goes well, the field will have to navigate ethical and political obstacles. Since 1996, Congress has attached a rider to the Health and Human Services appropriations bill that precludes federal funding of research in which human embryos are destroyed. The rider, attached again to the 2002 bill, which has yet to be signed into law, defines embryos to include parthenotes.

Privately funded research is not affected by that legislation, and some believe that federal funds should be allowed for research on parthenotes that have no potential to grow into people. After all, said Hall, even a skin cell has the potential to become a viable embryo through cloning technology. "If you look at it that way, then a body cell may have more rights or moral basis than a parthenote."

For society to decide what kinds of human entities deserve

rights, said Charo, it must consider why rights exist and why society recognizes them. One view, she said, is that rights accrue only if they have some value to the rights holder, leaving embryonic entities with no rights if they lack any potential for life.

Many critics of embryo research don't see it that way, said John Eppig, a researcher at Jackson Laboratory in Bar Harbor, Maine, who sat on a 1995 federal commission that gathered public comments on human embryo research. "The same people who were up in arms about doing research on embryos were up in arms about research on parthenotes," he said. "They correlated this with virgin birth. They correlated it with Christ."

Douglas Johnson of the National Right to Life Committee said he was unconvinced that parthenotes deserve no protection. "With respect to any of these things created with human genetic material," he said, "the burden of proof must rest with those who argue they are not human and that it's okay to kill them."

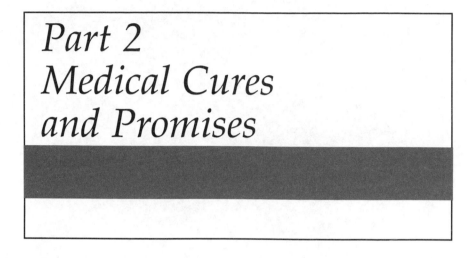

Part 2
Medical Cures
and Promises

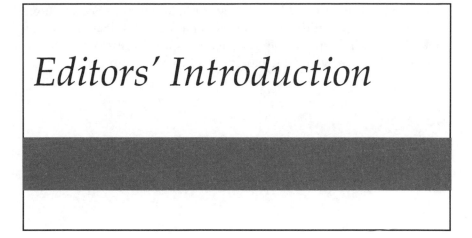

Editors' Introduction

Will Superman ever walk again? One of the most tragic stories of the last decade was that of Christopher Reeve, the actor who played Superman in the recent movies. He fell from his horse and now, from being a truly magnificent specimen of humankind, he is paralyzed from the neck down. One of the most heartwarming stories of the last decade has been the way in which Reeve has fought back, not only taking on acting roles but in pushing publically for research into spinal injuries and related afflictions. Who could fault him if he put his hopes into stem cell research? Nor, according to the news reports with which we begin this section, are such hopes misplaced. Even if not immediately, it now seems that real advances are truly being made in the quest for cures of spinal injuries—and indeed of many other illnesses and diseases that afflict human beings. From the brain to the spine to the blood and more. To give you some idea of the full scope and potentiality of stem cell research, we turn to the NBAC report to President Clinton, offering an extract detailing some of the possible applications of such investigations. You will see just how broad the range seems to be. Then, to conclude this section, taken from *Science*, there is an in-depth look at the way in which Parkinson's disease is being tackled, with information about successes and failures and of the hopes for tomorrow and the barriers still to be surmounted.

6

Paralyzed Mouse Walks Again as Scientists Fight Stem Cell Ban

Katty Kay and Mark Henderson

A video showing mice that have been partially cured of paralysis by injections of human stem cells was released last night by American scientists. They are seeking to head off a ban on government funding of similar research.

Researchers at Johns Hopkins University in Baltimore broke with standard scientific practice to screen the tape before details of their research have been formally published, in the hope that it will convince President Bush of the value of stem cell technology.

The U.S. government is considering whether to outlaw all federal funding of studies using stem cells taken from human embryos, which promise to provide new treatments for many conditions, including paralysis and Parkinson's disease.

Opponents argue that the research is immoral as the cells are taken from viable human embryos. President Bush has suspended federal funding of such work and has announced a review of its future. He was urged this week by the Pope to outlaw the practice.

John Gearhart and Douglas Kerr, who led the privately funded research, hope that the tape will have a decisive impact on the debate by showing the potential of the technique. It shows mice paralyzed by motor neuron disease once again able to move their limbs, bear their own weight, and even move around after injections of human embryonic stem cells in their spinal cords.

Dr. Kerr said that the team hopes to start human clinical trials within three years but that a federal funding ban would deal a "potentially fatal blow" to its efforts.

Details of its research were first revealed in November last year, though it has yet to be published in a peer-reviewed journal. In this case, however, the team took the decision to show the tape to Tommy Thompson, the U.S. Health and Human Services secretary, who is conducting a review of stem cell funding for President Bush, and to Pete Domenici, a Republican senator. It is now to be released to the public as well.

Medical research charities said the video would have a major impact. "I wish the president would see this tape," said Michael Manganiello, vice president of the Christopher Reeve Paralysis Foundation, named after the Superman actor who was paralyzed in a riding accident.

"When you see a rat going from dragging his hind legs to walking, it's not that big a leap to look at Christopher Reeve, and think how this might help him," he said.

In the experiment, 120 mice and rats were infected with a virus that caused spinal damage similar to that from motor neuron disease, the debilitating condition that affects Professor Stephen Hawking. The disease is generally incurable and sufferers usually die from it within two to six years.

When fluid containing human embryonic stem cells was infused into the spinal fluid of the paralyzed rodents, every one of the animals regained at least some movement. In previous tests, stem cells have been transplanted directly into the spinal cord. Infusing the fluid is far less invasive and would make eventual treatment in humans much easier.

Dr. Kerr said the limited movement seen was a reflection of the limited research, not of the limits to stem cells themselves.

"I would be a fool to say that the ceiling we have now is the same ceiling we'll see in two years," he said. "We will be smarter and the stem cell research even more developed."

However, the prospect of human trials in three years depends on the outcome of a political and ethical debate over whether the U.S. government will allow federal funding for stem cell research. If President Bush decides not to approve government funds for research, that would set the timetable back ten to twelve years for tests in humans, said Dr. Kerr.

The controversy stems from the fact that human embryos must be destroyed in order to retrieve the stem cells. Mr. Bush is under pressure from conservative Republicans and Roman Catholics not to back the research on moral grounds.

Some top American scientists, who are becoming increasingly frustrated with the funding limitations, have left for Britain where government funding is available. The British government has approved stem cell research on the ground that it could help to cure intractable disease.

The research on rodents at Johns Hopkins took stem cells from five- to nine-week-old human fetuses that had been electively aborted.

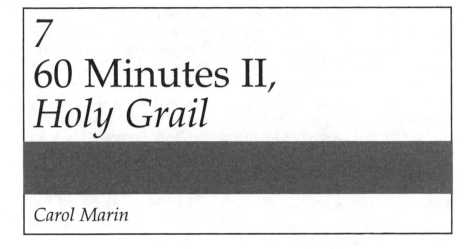

7
60 Minutes II,
Holy Grail

Carol Marin

Stem cells are thought of as the Holy Grail of medicine. One young boy agrees with that. He made medical history because he's been cured of his life-threatening disease. The key to his cure did not come from a human embryo, where all the controversy is, but from something that is routinely tossed in the garbage—an umbilical cord. Umbilical cords were always considered medical waste. Not anymore.

That's why new parents like Pam Dorne and Stephen Ayers of suburban Chicago have decided to save their children's umbilical cord blood. Dorne gave birth last spring to a baby boy, Kyle.

After a baby is born, there is just a fifteen-minute window to retrieve the four to six ounces of blood in the umbilical cord. And in that blood are potentially life saving stem cells that can be saved for future use.

"This is really where, I think, so much of biomedicine is going to be going in the twenty-first century," says Dr. Andrew Yeager of the University of Pittsburgh.

For instance, when stem cells from umbilical cord blood are injected into a person's vein, they migrate to the bone marrow and can create what Dr. Yeager calls a blood factory, replacing diseased blood with healthy blood. According to the National Institutes of Health, stem cells may one day be able repair the body's tissue and muscle and cure everything from spinal cord injuries to Alzheimer's.

"It's not just pie-in-the-sky speculation," says Yeager. "There are studies that would suggest that other organ dysfunction—nerve damage, heart damage, brain-cell damage—might actually be fixed."

It has the potential to make paralyzed patients walk and make Alzheimer's sufferers remember.

That potential is what Dr. Yeager was counting on to cure a young patient named Keone Penn.

Keone suffers from a case of sickle cell, a painful genetic blood disease. He was diagnosed when he was six months old. He was five when his sickle cell caused a stroke.

"All I remember is I woke up and my mama was beside me and there was a basket beside me and a teddy bear," he says. "It was very scary, I mean, whew."

For six years, Keone and his mother, Leslie Penn, were constantly in and out of an Atlanta hospital to receive transfusions to stave off another potentially deadly stroke. By the time Keone was eleven, the transfusions were becoming less effective and he had excruciating pain in his joints and lower back.

"The pain is usually so intense that even morphine, Demerol, those heavy-duty medicines don't really touch it," Leslie Penn says. "All you can really do is pray that he'll just go to sleep."

Keone says he's tough, but at fifteen, he looks much younger. Sickle cell stunted his growth—he's just four feet, nine inches, tall—and restricted what he could do.

"I was impaired from doing a lot of things that normal kids do, like sports or anything or run," he says. "Couldn't play basketball. Because, you know, some people like roughhousing when they play basketball and they can knock you over and push you and that could really hurt me."

The odds were that Keone had, at best, only five years to live. So Yeager decided to take a chance on a new procedure. Never before had stem cells from umbilical cord blood been used to treat sickle cell.

"The goal here is that these stem cells, which are in a relatively high proportion in cord blood—higher than they would be in our own bone marrow and definitely higher than in our own circulating blood—could then be injected and would take hold and again, make more of themselves. And make a whole new blood factory."

Yeager told the family he wasn't sure the procedure would work.

"He just basically said, 'This is just a fifty-fifty chance and it's up to you all if you want to do it, I can't offer you any guarantees,'" recalls Leslie Penn.

Keone Penn remembers how his mother told him: "She came in the room looking very depressed. Pulled the chair up, sat beside me in the bed, and told me everything. And I almost started to cry. But she was very calm about it. She told me everything, said, 'You got— you may—have five years to live,' you know."

Ordinarily, patients with a severe case of sickle cell, like Keone's, would have had a bone-marrow transplant. That's because until recently bone marrow was the only source for stem cells.

But bone marrow transplants can be tricky because there must be a precise match between the person donating the bone marrow and the patient receiving it. In Keone's case, no match could be found.

Stem cells from umbilical cord blood don't need an exact match.

Dr. Yeager and his team found a match that was close enough in a cryogenic tank at the New York Public Blood Bank, which since 1992 has slowly been collecting donations of umbilical cord blood.

Over Christmas vacation of 1998, after intensive chemotherapy to destroy Keone's bad blood, he was injected with the stem cells.

After a few weeks, something extraordinary happened—the stem cells changed his entire blood system from type O to type B.

"That concept there is the one that really blows my mind," says Leslie Penn. "The thought that your whole blood type is changed. The umbilical cord cell's donor, he took on their blood type."

A year later, doctors declared that the sickle cells in Keone's body had disappeared. Today, he is considered cured.

It was umbilical cord stem cells that cured Keone, not stem cells from human embryos. While the use of embryonic stem cells has generated fierce controversy, umbilical cord stem cells have attracted little attention and no political debate. And now it seems, more and more new parents have decided to bank their hopes on the stem cells in their newborn's cord blood.

Moments after Pam Dorne gave birth to baby Kyle, his cord blood was sealed, packed in dry ice, and given to a courier. Within hours, the package was on a plane bound for Tucson, Arizona, where the largest privately run cord blood bank in the country is located.

There, a child's umbilical-cord blood is stored in a cryogenic tank at a temperature of minus 400 degrees Fahrenheit.

Dr. David Harris, laboratory director of the Cord Blood Registry, says it takes only a small vial of cord blood to change a person's entire system.

So far, Cord Blood Registry has collected about thirty thousand samples from families willing to pay a $1,300 flat fee and $95 a year to analyze and privately store their baby's cord blood. The company has taken in over $40 million so far, selling a kind of biological insurance.

"Part of the issue when people bank," says Harris, "is because they have a family history or they work or live in a place where there is a potential for cancer. But part of it is for peace of mind."

According to the American Academy of Pediatrics, that peace of mind isn't worth the money. The academy says the chances a family will ever need to use its frozen cord blood are very small. What they say makes more sense is to donate cord blood to a public bank, the kind where Keone Penn got his stem cells.

That is something Pam Dorne, an obstetrician, says she understands in her professional life. But her own personal choice was a private bank, she says, for one reason.

"If the American Academy of Pediatrics could tell me that none of my children would ever have a problem," she says, "or that if they had a problem, I would be guaranteed that there would be enough donors and somebody would match them, that would be perfectly reasonable. But I don't think anybody has that crystal ball."

What saved Keone Penn's life, Dr. Yeager says, is a public blood bank and the umbilical cord blood from an anonymous donor.

"If they wish to pay, that's absolutely fine," he says of patients. "But to look at a larger, greatest good for greatest number, I would contend that a volunteer donation to a public blood bank would make the most sense."

Meanwhile, Keone, a pioneer, is doing things he's never done before.

"I discovered the other day that I like playing basketball," Keone says. "I never played basketball, 'cause I've always been disabled to play it and to have fun."

Keone, who one day hopes to become a chef, still has some major health problems as a result of infections that occur in most stem-cell transplants. Because of steroids and other medication, he has arthritis, walks with a limp, and will need joint replacement in his hips and knee. But the good news is the sickle cell that was killing him is gone.

"I love stem cells," he says. "I mean they saved my life. If it weren't for them I wouldn't, you never know, I probably wouldn't be here today."

Keone doesn't know where the cord blood came from or who is the owner. He says he would like to know, just so he could say, "Thank You."

8

Human Stem Cell Research and the Potential for Clinical Application

National Bioethics Advisory Commission
Potential Medical Applications

POTENTIAL MEDICAL APPLICATIONS OF HUMAN ES CELL AND EG CELL RESEARCH

Although research into the use of ES and EG cells is still at an early stage, researchers hope to make a contribution to disease treatment in a variety of areas. The ability to elucidate the mechanisms that control cell differentiation is, at the most elemental level, the promise of human ES and EG cell research. This knowledge will facilitate the efficient, directed differentiation of stem cells to specific cell types. The standardized production of large, purified populations of human cells such as cardiomyocytes and neurons, for example, could provide a substantial source of cells for drug discovery and transplantation therapies.[1] Many diseases, such as Parkinson's disease and juvenile-onset diabetes mellitus, result from the death or dysfunction of just one or a few cell types, and the replacement of those cells could offer effective treatment and even cures.

Reprinted from the National Bioethics Advisory Commission (NBAC).

Substantial advances in basic cell and developmental biology are required before it will be possible to direct human ES cells to lineages of human clinical importance. However, progress has already been made in the differentiation of mouse ES cells to neurons, hematopoietic cells, and cardiac muscle.[2] Human ES and EG cells could be put to use in targeting neurodegenerative disorders, diabetes, spinal cord injury, and hematopoietic repopulation, the current treatments for which are either incomplete or create additional complications for those who suffer from them.

USE OF HUMAN ES CELLS AND EG CELLS IN TRANSPLANTATION

One of the major causes of organ transplantation and graft failure is immune rejection, and a likely application of human ES and EG cell research is in the area of transplantation. Although much research remains to be done, ES cells derived through SCNT [somatic cell nuclear transfer] offer the possibility that therapies could be developed from a patient's own cells. In other words, a patient's somatic cells could be fused with an enucleated oocyte and developed to the blastocyst stage, at which point ES cells could be derived for the development of cell-based therapy. This essentially is an autologous transfer. Thus, issues of tissue rejection due to the recognition of foreign proteins by the immune system are avoided entirely. In addition, research to establish xenotransplantation (i.e., interspecies transplantation) as a safe and effective alternative to human organ transplantation is still in its infancy. Alternately, other techniques that would be immunologically compatible for transplantation purposes could be used to generate stem cells, such as

1. banking of multiple cell lines representing a spectrum of major histocompatibility complex (MHC) alleles to serve as a source for MHC matching, and/or
2. creating universal donor lines, in which the MHC genes could be genetically altered so rejection would not occur, an approach that has been tried in the mouse with moderate success.[3]

Autologous transplants would obviate the need for immunosuppressive agents in transplantation as it would decrease a central danger to transplant patients—susceptibility to other diseases. Autologous transplants might address problems ranging from the supply of donor organs to the difficulty of finding matches between donors and recipients. Research on ES cells could lead to cures for diseases that require treatment through transplantation, including autoimmune diseases such as multiple sclerosis, rheumatoid arthritis, and systemic lupus erythematosus. These cells also might hold promise for treating type-I diabetes,[4] which would involve the transplantation of pancreatic islet cells or beta cells produced from autologous ES cells. These cells would enter the pancreas and provide normal insulin production by replacing the failing resident islet cells.

STUDIES OF HUMAN REPRODUCTION AND DEVELOPMENTAL BIOLOGY

Research using human ES and EG cells could offer insights into developmental events that cannot be studied directly in the intact human embryo but that have important consequences in clinical areas, including birth defects, infertility, and pregnancy loss.[5] ES and EG cells provide large quantities of homogeneous material that can be used for biochemical analysis of the patterns of gene expression and the molecular mechanisms of embryonic differentiation.

CANCER THERAPY

Human ES and EG cells may be used to reduce the tissue toxicity brought on by cancer therapy.[6] Already, bone marrow stem cells, representing a more committed stem cell, are used to treat patients after high dose chemotherapy. However, the recovered blood cells appear limited in their ability to recognize abnormal cells, such as cancer cells. It is possible that injections of ES and EG cells would revive the complete immune response to patients undergoing bone marrow transplantation. Current approaches aimed at manipulating the immune system after high-dose chemotherapy so that it recognizes cancer cells specifically have not yet been successful.

DISEASES OF THE NERVOUS SYSTEM

Some believe that in no other area of medicine are the potential benefits of ES and EG cell research greater than in diseases of the nervous system.[7] The most obvious reason is that so many of these diseases result from the loss of nerve cells, and mature nerve cells cannot divide to replace those that are lost. For example, in Parkinson's disease, nerve cells that make the chemical dopamine die; in Alzheimer's disease, it is the cells that make acetylcholine that die; in Huntington's disease the cells that make gamma aminobutyric acid die; in multiple sclerosis, cells that make myelin die; and in amyotrophic lateral sclerosis, the motor nerve cells that activate muscles die. In stroke, brain trauma, spinal cord injury, and cerebral palsy and mental retardation, numerous types of cells are lost with no built-in mechanism for replacing them.

Preliminary results from fetal tissue transplantation trials for Parkinson's disease suggest that supplying new cells to a structure as intricate as the brain can slow or stop disease progression.[8] Yet the difficulty of obtaining enough cells of the right type—that is, dopamine-producing nerve cells—limits the application of this therapy. In 1999, scientists developed methods in animal models to isolate dopamine precursor cells from the dopamine-producing region of the brain and coax them to proliferate for several generations in cell culture. When these cells were implanted into the brains of rodents with experimental Parkinson's disease, the animals showed improvements in their movement control.[9] Scientists also have learned to instruct a stem cell from even a nondopamine region to make dopamine.[10] A large supply of "dopamine-competent" stem cells, such as ES cell lines, could remove the barrier of limited amounts of tissue. (See Exhibit 2-C.)

Another recent development eventually may provide treatments for multiple sclerosis and other diseases that attack the myelin coating of nerves. Scientists have successfully generated glial cells that produce myelin from mouse ES cells.[11] When these ES cell-derived glial cells were transplanted in a rat model of human myelin disease, they were able to interact with host neurons and efficiently myelinate axons in the rat's brain and spinal cord.[12]

Other diseases that might benefit from similar types of approaches include spinal cord injury, epilepsy, stroke, Tay-Sachs disease, and pediatric causes of cerebral palsy and mental retardation. In mice, neural stem cells already have been shown to be effective in replacing cells throughout the brain and in some cases are capable of correcting neurological defects.[13] Human neural stem cells also have recently been isolated and have been shown to be responsive to developmental signals and to be willing to replace neurons when transplanted into mice.[14] These recent discoveries of ways to generate specific types of neural cells from ES cells hold much promise for the treatment of severe neurological disorders that today have no known cure.

Exhibit 2-C: Potential Treatment for Parkinson's Disease

Parkinson's disease is a degenerative brain disease that affects 2 percent of the population over age seventy. Symptoms include slow and stiff movements, problems with balance and walking, and tremor. In more advanced cases, the patient has a fixed, staring expression, walks with a stooped posture and short, shuffling pace, and has difficulty initiating voluntary movements. Falls, difficulty swallowing, incontinence, and dementia may occur in the late stages. Patients often lose the ability to care for themselves and may become bedridden.

The cause of this illness is a deficiency of the neurotransmitter dopamine in specific areas of the brain. Treatment with drugs such as levodopa often is effective in relieving the symptoms. However, as the disease progresses, treatment often becomes more problematic, with irregular responses, difficulty adjusting doses, and the development of side effects such as involuntary writhing movements. Brain surgery with transplantation of human fetal tissue has shown promise as therapy.

Stem cell transplantation also may be a promising therapy for Parkinson's disease. The injection of stem cells that can differentiate into brain cells may offer a means of replenishing neurons that are capable of synthesizing the deficient neurotransmitter. It is possible that stem cell transplantation may be simpler and more readily available than fetal tissue transplantation.

DISEASES OF THE BONE AND CARTILAGE

Because ES and EG cells constitute a relatively self-renewing population of cells, they can be cultured to generate greater numbers of bone or cartilage cells than could be obtained from a tissue sample. If a self-renewing, but controlled, population of stem cells can be established in a transplant recipient, it could effect long-term correction of many diseases and degenerative conditions in which bone or cartilage cells are deficient in numbers or defective in function. This could be done either by transplanting ES and EG cells to a recipient or by genetically modifying a person's own stem cells and returning them to the marrow. Such approaches hold promise for the treatment of genetic disorders of bone and cartilage, such as osteogenesis imperfecta and the various chondrodysplasias. In a somewhat different potential application, stem cells perhaps could be stimulated in culture to develop into either bone- or cartilage-producing cells. These cells could then be introduced into the damaged areas of joint cartilage in cases of osteoarthritis or into large gaps in bone that can arise from fractures or surgery. This sort of repair would have a number of advantages over the current practice of tissue grafting.[15]

BLOOD DISORDERS

The globin proteins are essential for transport of oxygen in the blood, with different globins expressed at different developmental stages. The epsilon globin gene is expressed only in embryonic red blood cells. When this gene—which is not normally expressed in the adult—is artificially turned on in sickle cell patients, it blocks the sickling of the cells that contain sickle cell hemoglobin. Research involving ES cells could help answer questions about how to turn on the epsilon globin gene in adult blood cells and thereby halt the disease process. Stem cell research also may help produce transplantable cells that would not contain the sickle cell mutation.

TOXICITY AND DRUG TESTING

Human stem cell research offers promise for use in testing the beneficial and toxic effects of biologicals, chemicals, and drugs in the most relevant species for clinical validity—humans. Such studies could lead to fewer, less costly, and better designed human clinical trials yielding more specific diagnostic procedures and more effective systemic therapies. Beyond the drug development screening of pharmacological agents for toxicity and/or efficacy, human stem cell research could define new research approaches for clarifying the complex association of environmental agents with human disease processes.[16] It also makes possible a new means of conducting detailed investigations of the underlying mechanisms of the effects of environmental toxins or mixtures of toxins, including their subtle effects on the developing embryonic and fetal development tissue systems.

TRANSPLANTABLE ORGANS

Several researchers are investigating ways to isolate AS cells and create transplantable organs that may be used to treat a multitude of diseases that do not rely upon the use of embryonic or fetal tissue. Moreover, if it is found to be possible to differentiate ES cells into specific cell types, such stem cells could be an important source of cells for organ growth. For example, recent developments in animals have shown that it may be possible to create entire transplantable organs from a tissue base in a manner that would overcome such problems as the limited supply of organs and tissue rejection. Such a development—producing this tissue base by directing the growth of human embryonic cells—could be a major breakthrough in the field of whole organ transplantation.

For example, using tissue engineering methods, researchers have successfully grown bladders in the laboratory, implanted them into dogs, and shown them to be functional.[17] To create the bladders, small biopsies of tissue were taken from dog bladders. The biopsied tissue was then teased apart to isolate the urothelial tissue and muscle tissue, which were then grown separately in culture.[18] The tissue was then applied to a mold of biodegradable material with the urothelial tissue on the inside and the muscle tissue on the outside. The new organs were transplanted within five weeks.[19]

Dogs that received the tissue-engineered organs regained 95 percent of their original bladder capacity, were continent, and voided normally. When the new organs were examined eleven months later, they were completely covered with urothelial and muscle tissue and had both nerve and blood vessel growth. Dogs that did not undergo reconstructive procedures or only received implants of the biodegradable molds did not regain normal bladder function.[20] This accomplishment marks the first time a mammalian organ has been grown in a laboratory. The ability to create new organs by seeding molds with cells of specific tissue types would be extremely useful in treating children with congenital malformations of organs and people who have lost organs due to trauma or disease.[21]

SUMMARY

Currently, human ES cells can be derived from the inner cell mass of a blastocyst (those cells within the conceptus that form the embryo proper), and EG cells can be derived from the primordial germ cells of fetuses. These cells, present in the earliest stages of embryo and fetal development, can generate all of the human cell types and are capable, at least for some time, of self-renewal. A relatively renewable tissue culture source of human cells that can be used to generate a wide variety of cell types would have broad applications in basic research, transplantation, and other important therapies, and a major step in realizing this goal was taken in 1998 with the demonstration that human ES and EG cells can be grown in culture. The clinical potential for these stem cells is vast—they will be important for *in vitro* studies of normal human embryogenesis, human gene discovery, and drug and teratogen testing and as a renewable source of cells for tissue transplantation, cell replacement, and gene therapies.

NOTES

1. J. A. Thomson et al., "Embryonic Stem Cell Lines Derived from Human Blastocysts," *Science* 282 (1998): 1145–47.
2. O. Brustle et al., "*In vitro*-generated Neural Precursors Participate in Mammalian Brain Development," *Proceedings of the National Academy of Sciences USA* 94 (1997): 14809–14.

3. National Institute of Allergy and Infectious Diseases (NIAID), "What Would You Hope to Achieve from Stem Cell Research?" Response to Senator Specter's Inquiry, March 25, 1999.

4. Senate Appropriations Subcommittee on Labor, Health, and Human Services, January 12, 1999, D. Melton, *Testimony of the Juvenile Diabetes Foundaton*; Senate Appropriations Subcommittee on Labor, Health, and Human Services, December 2, 1998, H. Varmus, *Testimony*.

5. Thomson et al., "Embryonic Stem Cell Lines."

6. National Cancer Institute (NCI), "What Would You Hope to Achieve from Stem Cell Research?" Response to Senator Specter's Inquiry, March 25, 1999.

7. J. D. Gearhart, "New Potential for Human Embryonic Stem Cells," *Science* 282 (1998): 1061–61; Varmus, *Testimony*.

8. C. R. Freed et al., "Double-Blind Controlled Trial of Human Embryonic Dopamine Cell Transplants in Advanced Parkinson's Disease: Study Design, Surgical Strategy, Patient Demographics, and Pathological Outcome" (presented to the American Academy of Neurology, April 21, 1999).

9. National Institute of Neurological Disorders and Stroke (NINDS), "What Would You Hope to Achieve from Stem Cell Research?" Response to Senator Specter's Inquiry, March 25, 1999.

10. J. Wagner et al., "Induction of a Midbrain Dopaminergic Phenotype in *Nurr1*-Overexpressing Neural Stem Cells by Type 1 Astrocytes," *Nature Biotechnology* 17 (1999): 653–59.

11. O. Brustle et al., "Embryonic Stem Cell–Derived Glial Precursors: A Source of Myelinating Transplants," *Science* 285 (1999): 754–56.

12. Ibid.

13. H. D. Lacorazza et al., "Expression of Human b-hexosaminidase a Subunit Gene (the Gene Defect of Tay-Sachs Disease) in Mouse Brains upon Engraftment of Transduced Progenitor Cells," *Nature Medicine* 2 (1996): 424–29; C. M. Rosario et al., "Differentiation of Engrafted Multipotent Neural Progenitors towards Replacement of Missing Granule Neurons in Meander Tail Cerebellum May Help Determine the Locus of Mutant Gene Action," *Development* 124 (1997): 4213–24; E. Y. Snyder et al., "Multipotent Neural Precursors Can Differentiate toward Replacement of Neurons Undergoing Targeted Apoptotic Degeneration in Adult Mouse Neocortex," *Proceedings of the National Academy of Sciences USA* 94 (1997): 11663–68; E. Y. Snyder, R. M. Taylor, and J. H. Wolfe, "Neural Progenitor Cell Engraftment Corrects Lysosomal Storage Throughout the MPS VII Mouse Brain, *Nature* 374 (1995): 367–70; B. D. Yandava, L. L. Billinghurst, and E. Y. Snyder, "'Global' Cell Replacement Is Feasible via Neural Stem Cell Transplantation: Evidence from the Dysmyelinated *Shiverer* Mouse Brain," *Proceedings of the National Academy of Sciences USA* 96 (1999): 7029–34.

14. J. D. Flax et al., "Engraftable Human Neural Stem Cells Respond to

Developmental Cues, Replace Neurons, and Express Foreign Genes," *Nature Biotechnology* 16 (1998): 1033–39.

15. National Institute of Arthritis and Musculoskeletal and Skin Diseases (NIAMS), "What Would You Hope to Achieve from Stem Cell Research?" Response to Senator Specter's Inquiry, March 25, 1999.

16. National Institute of Environmental Health Sciences (NIEHS), "What Would You Hope to Achieve from Stem Cell Research?" Response to Senator Specter's Inquiry, March 25, 1999.

17. F. Oberpenning et al., "De Novo Reconstruction of a Functional Mammalian Urinary Bladder by Tissue Engineering," *Nature Biotechnology* 17 (1999): 149–55.

18. J. H. Tanne, "Researchers Implant Tissue-Engineered Bladders," *British Medical Journal* 318 (1999): 350B.

19. Ibid.

20. Oberpenning et al., "De Novo Reconstruction."

21. Tanne, "Researchers Implant Tissue-Engineered Bladders."

9

Fetal Neuron Grafts Pave the Way for Stem Cell Therapies

A decade of experimental treatments using fetal neurons to replace brain cells that die in Parkinson's disease can provide lessons for planning stem cell therapies.

Marcia Barinaga

Swedish neuroscientist Anders Björklund and his colleagues may have caught a glimpse of what the future holds for the treatment of failing organs. For more than ten years, Björklund has been part of a team at Lund University in Sweden that has been grafting neurons from aborted fetuses into the brains of patients with Parkinson's disease. In many cases, the transplanted cells have dramatically relieved the patients' symptoms, which include slowness of movement and rigidity. That is just the kind of therapy that stem cell researchers hope to make routine for treating all sorts of degenerative diseases, if they can coax the cells to develop into limitless supplies of specific cell types that can be used to repair or replace damaged organs.

Although the current Parkinson's treatment uses fetal cells that have already developed into a particular type of neuron, the promising results represent a "proof of principle that cell replacement actually works," says Björklund. The results have given researchers increased confidence that, if they can manipulate stem cells to

Reprinted with permission from *Science* 287, no. 5457 (2000): 1421–22.

develop into the kind of neuron the Lund group and others are using—a big challenge—the new cells would take over the work of damaged cells in the brains of Parkinson's patients. If so, Parkinson's treatment could be among the first applications of stem cell therapy.

The successes have also increased the urgency of developing stem cell treatments, because despite their promise, there are many reasons that fetal cells will never be widely used to treat Parkinson's disease. The reasons range from ethical concerns, such as the protests by antiabortionists that led the governor of Nebraska to urge that research involving fetal tissue be shut down at the University of Nebraska,[1] to the fact that there will never be enough fetal tissue to treat all the people who need it. Parkinson's disease afflicts 1 million people in the United States alone.

Researchers are now looking closely at the results from fetal cell transplants for lessons that will help guide future work with stem cells. There are still many hurdles to overcome, but this first round of cell replacement in the brain sets a "gold standard" that stem cells must meet if they are to become the basis for new Parkinson's treatments, says neuroscientist and stem cell researcher Evan Snyder of Harvard Medical School in Boston.

Parkinson's disease is a logical candidate for cell replacement therapy, in part because conventional treatments have had limited success. The disease is caused by the death, for unknown reasons, of a particular group of brain neurons that produce dopamine, one of the chemicals that transmit signals between nerve cells. Afflicted people lose the ability to control their movements, ultimately becoming rigid. Treatment with levodopa (L-dopa), a drug that is converted to dopamine by the brain, alleviates these symptoms, but as the neurons continue to die, L-dopa's effectiveness wanes. Researchers first tried replacing the dopamine-producing cells by grafting into the affected region cells from the adrenal medulla gland. These cells are not neurons, but they make dopamine and can be coaxed to become neuronlike. The treatment reversed Parkinson's symptoms in rats, but produced little lasting improvement in human patients, probably because the cells died or stopped making dopamine, says John Sladek, chair of neuroscience at the Chicago Medical School.

Researchers have had better luck grafting immature neurons

taken from aborted human fetuses. Dozens of patients who have received these experimental dopamine-neuron grafts over the past ten years have had up to a 50 percent reduction in their symptoms. And the effects appear to last. Using positron emission tomography to image the brain, Olle Lindvall of Lund University and a team of colleagues in Lund and at Hammersmith Hospital in London reported in the December issue of *Nature Neuroscience* that, in one patient, the transplanted neurons are still alive and making dopamine ten years after the surgery. That's encouraging, says neurotransplant researcher Ole Isacson of Harvard Medical School: Whatever killed the brain's own dopamine-producing neurons doesn't seem to have killed the transplanted cells.

Still, fetal cell transplants are plagued by problems that can never be overcome. Aside from ethical concerns about scavenging neurons from aborted fetuses, there are practical issues. It takes six fetuses to provide enough material to treat one Parkinson's patient, in part because as many as 90 to 95 percent of the neurons die shortly after they are grafted. Indeed, Lund's Björklund says, the cell supply is so limited that researchers have not even been able to test some possible avenues for fetal cell transplants. The neurons that die in Parkinson's originate in a brain region called the substantia nigra and send their long axons to several other areas, where they release dopamine. So far, researchers have put the cell grafts into only one of these areas, the putamen—and even there, they have not yet transplanted enough neurons to restore normal dopamine levels in most cases.

Even if researchers can develop techniques that diminish the fetal cell die-off, there will never be enough fetuses available to make this an "everyday procedure," says Sladek. What's more, the brain material recovered from aborted fetuses "comes out in a form that makes it difficult to . . . standardize" in terms of quality and purity, Björklund says. This is likely why some patients do far better than others—uncertainty that would be unacceptable in a standard medical treatment.

Consequently, researchers are pinning their hopes on cultured stem cells. They would eliminate a continuing dependence on aborted fetuses, although the ethical concerns won't be completely laid to rest unless researchers can use stem cells obtained from adults rather than embryos.[2] And the supply of cultured cells could be unlimited, allowing tests of grafts into the putamen and possibly into other brain areas as well. The cell treatment, moreover, could be standardized and controlled to assure a more predictable outcome. "The ability to grow the cells of interest will make this a routine technology," predicts neuroscientist Ron McKay, whose team is working on ways to culture neural stem cells at the National Institute of Neurological Disorders and Stroke.

But to make this brave new world of cell replacement technology a reality, researchers must first learn how to keep stem cells dividing for many generations in culture and then be able to trigger them to differentiate into the type of neuron they want. Stem cells presumably have the ability to differentiate into any of the several different types of dopamine neurons the brain contains, but it may be crucial to use the specific type of dopamine neurons that die in Parkinson's. Researchers doing the fetal cell transplants specifically select these neurons—known as nigral neurons because they originate in the substantia nigra—when they harvest neurons for grafting from fetal brains. Nigral neurons are "genetically programmed and designed to be a dopamine neuron in the appropriate brain circuit," Sladek says.

Among other things, nigral neurons may respond better to local conditions, producing just enough dopamine. Experience with L-dopa treatment has shown that too much of the neurotransmitter can be just as problematic as too little, causing uncontrollable, jerky movements in patients. Researchers worry that stem cells coaxed to develop into dopamine neurons may become one of the nonnigral types and will not regulate their dopamine output in the appropriate ways. "It is like putting the right alternator into your car," Sladek says. "If you put in one designed for another model of car, it may not work as well."

Getting stem cells to differentiate into the right type of neuron may be only part of the problem. Neurons in their natural environment are surrounded by support cells called "glia," which nurture the neurons and even modulate their activity, and optimal cell transplants may require replacing not only dopamine neurons but also the glia that normally surround them, Harvard's Snyder suggests.

But some researchers believe that the brain itself may be able to overcome the hurdles of producing both the proper neurons and the support cells they need. Snyder's lab has shown in animal experiments that stem cells put into the brain can be influenced by the brain environment to differentiate into both neurons and support cells. He envisions someday putting stem cells into Parkinson's brains and letting the brain tell them which cell types to become.

Even if it turns out not to be quite that simple, Parkinson's poses a much less daunting challenge for cell replacement therapy than do other neurological disorders. "The dopaminergic system is a fairly easy system to work with compared to sensorimotor or visual systems or spinal cord," says Isacson. That, he says, is because the nigral neurons lost in Parkinson's disease have a diffuse and relatively nonspecific network of connections in the brain areas they link up to, rather than the very intricate and precise connections made by neurons in many other parts of the nervous system.

The treatment of most other brain disorders would likely require coaxing new neurons to make very precise connections, a task that no one is sure how to achieve. But in the case of Parkinson's disease, simply getting neurons to release dopamine in the correct general area helps patients. Because of that difference, Isacson predicts that "it will take some time to get other diseases to benefit from all these discoveries." Nevertheless, a successful Parkinson's treatment based on stem cells would still be a dramatic achievement. "You would help a huge number of patients," he says, "as many as the surgeons could do."

NOTES

1. E. Marshall, "Antiabortion Groups Target Neuroscience Study at Nebraska," *Science* 287 (2000): 202.

2. G. Vogel, "Can Old Cells Learn New Tricks?" *Science* 287 (February 25, 2000): 1418.

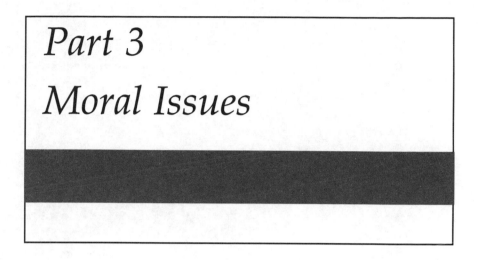

Part 3
Moral Issues

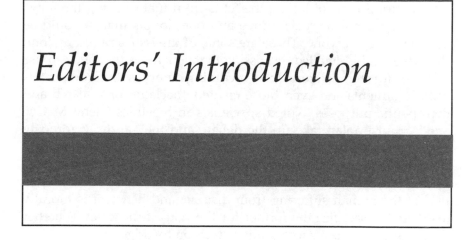

Editors' Introduction

Some people simply think that stem cell research is wrong, absolutely and completely. The conservative commentator Michael Novak is one. He takes a Kantian position, arguing that one should never destroy a human being for the benefit of another, although note that he and those whom he mentions favorably think that the notion of human being goes right back to the very moment of conception or its equivalent. Others would obviously disagree, including the ethicists drawn on by the biotech company Geron, who argue that in properly controlled circumstances one can indeed move forward ethically on stem cell research. Note that like many consequentialists, these people are concerned about overall good, basing their recommendations at least in part on global concerns. For them, part of the ethical justification of such research comes from the benefits that would (or should) be received by the many and not just the few. Also note that to some extent they try to side step the issue of the status of the fetus by arguing that if the fetus is already available (that is through abortion for unrelated reasons), then this opens the way for its use. In other words, they try to brick off the moral issue of killing the fetus from that of using the fetus. Obviously, this is a crucial move and one doubts that it would be acceptable to everyone. These are some of the issues raised by John Robertson in his detailed discussion that concludes that one can use embryos for stem cell research—including embryos discarded in fertility treatments and even those created specially for research and therapeutic purposes. Noted specialists in bioethics Glenn McGee and Arthur Caplan take up the debate and they argue vigorously that too much discussion of these issues has been based on politics first and ethics second, whereas truly the order should be reversed. They argue that, on balance, the worth of the embryo does not equal that of the human suffering from disease; and hence, it is morally justifiable to sacrifice the former for the sake of the latter. Whether they make their case successfully is for you to judge.

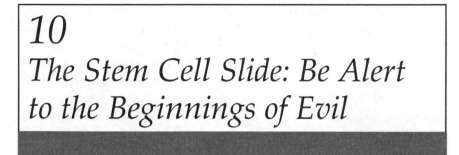

10

The Stem Cell Slide: Be Alert to the Beginnings of Evil

Michael Novak

I wish I could say that the president's speech to the nation on stem cells was as good as I had hoped. It was in many ways a wonderful speech: deeper and more philosophical than I have ever heard a president deliver, unusually balanced and fair in presenting opposing arguments, and clear in delineating both his own decision and the reason for it. It was, in addition, heartfelt and compassionate toward all families who have members suffering from awful diseases or disabilities. I can even see how the president convinced himself, at the end, that he had found a ray of daylight through the opposing arguments, and arrived at a moral decision that seemed to him sound, and also politically defensible. During the last few months, I have heard many persons who think they are very smart lay out their arguments on this question. Not one of them did as thorough, many-sided, fair, and clear a job as President Bush did in his speech.

At the end, though, my heart sank. The president tried to maintain a position of principle, but what he ended up doing, despite his best effort, was giving away the principle. He put the Full Faith and Credit of the U.S. government behind the principle of using human beings as a means, albeit for noble ends. He offered a reason for doing this: The stem cells for whose use in experimentation he commits federal resources come from embryos already destroyed. Why not bring good out of evil, he argues, by now using these stem cells, which will otherwise be wasted, to search for cures for awful diseases? The outcome is not certain, but at least it's noble to try. This is a lovely and tempting thought. The problem is that when the source of stem cells runs out—soon—then those on the other side will demand more stem cells from more embryos. The demand for usable stem cells will swell enormously. This is particularly true if good experimental results are obtained. But it will even be true if they aren't: Look, partisans will say, you were too stingy, too narrow. You have ceded the principle, so now give up more of the specifics. The glittering utopia of science beckons just ahead.

Be alert to the beginnings of evil. It never comes under the appearance of evil, but always under the appearance of the beautiful, the promising, the idealistic, the pleasant. "Stop it in its beginnings," the ancient principle runs: the sooner, the easier.

Politically, the decision may play very well among a substantial majority. It is already clear that those on the left who you hope will attack it are attacking it, which will only reinforce those among right-to-lifers who accept the president's obvious good will, often deeply moving words, clearly articulated argument, and patent depth of feeling. But I deeply fear the immense battles that lie ahead, and the gathering of heartened foes, who will very quickly sniff out the weak point and pry its own inner logic with all their force. It will take almost superhuman strength now for the president to hold the new position he has moved into, having surrendered the strongest ground.

That ground was a philosophical one, not a theological one, a ground born of reason rather than of faith. One of its classic articulators was Immanuel Kant. The president himself alluded to it in his speech, in the line about not using human beings as means for even the noblest ends. You must never use a human being as a means for even the noblest of ends. To use stem cells obtained by killing living human beings in their embryonic stage is still using them as a means. It is not enough to say that the wicked deed has already been done—that the embryos have already been killed. The purpose of that killing was to obtain the stem cells. One ought not to implicate oneself in that process, not even for the noblest and most beautiful ends.

One especially ought not to implicate the United States, which Hannah Arendt once called humankind's noblest experiment. For this nation began its embryonic existence by declaring that it held to a fundamental truth about right to life endowed in us by our Creator. The whole world depends on our upholding that principle.

We human beings very easily reason ourselves into taking positions that end up having the most tragic of consequences, positions of which we would never have approved had we seen those consequences at the time. For the fruit of the tree of knowledge over yonder appears to be very sweet, and we feel sure that if we eat of it, then happy endings (fit for a god) will result. Those endings have always turned to sulphur in our cheeks.

The fatal mistake often comes as a result of unexamined moral sentiments: affects and feelings that serve as moral guideposts without submitting to interrogations by reason. "I feel comfortable with that," President Bush—like President Clinton, indeed like just about everybody else in this fair land—is wont to say. It is as if Americans were ashamed to say that they reached a considered intellectual judgment, independently of their feelings. "I feel comfortable with that" seems itself to be more comfortable than "That's what I've reasoned to."

And this should cause us great uneasiness, because very often in the moral life, our feelings and sentiments are horrible guides to right action. It sometimes feels like sheer Hell to have to do the right thing, and most terribly *un*comfortable. For instance: When individuals in Nazi Europe made the personal decision to join the Resistance, they often did so feeling the most awful dread, sick in their stomachs about the prospect of being hunted out like animals, tortured, and killed in barbaric ways (on meat hooks, say). They could not afford to listen to their feelings. (Considerations of just this sort led Karol Wojtyla to abandon the philosophy of the moral sentiments he learned from Max Scheler, and to search for a philosophy that drove deeper to intellectual principle and a strong moral will.)

One thing this debate is showing the nation is the difference between those who can grasp, and be swayed by, intellectual principles and those who need warmer, fuzzier comforts for their senses, imaginations, and sentiments. Thus, when Senator [Orrin] Hatch finds it easier to imagine life in the womb than in a clump of cells in a petri dish, he is not wrong about what is occurring in his imagination and sentiments. But those means of perception do not dig deep enough into the *what-IS-it?* of those cells. To do that, cold intellect must go to work, beyond the comfort zone of imagination and sentiments.

And those cells are a living human being—at a very early stage, to be sure, but unmistakably human: not a rat, or a cocker spaniel, or a fish. That's precisely what makes them so desirable to researchers.

Desirable: There's a key word. It is transparent, now, how hungry researchers are to get their hands on these embryos. And perhaps on the money and the fame that beckon just ahead. As Mr. Dooley reminds us, when a fella says, 'tain't the money, it's the principle—it's the money. Suddenly, even great scientists are forgetting the basic biology still being taught in the latest textbooks: that the first appearance of *human* beings is in the fertilized egg embryo. Suddenly, scientists are fudging: "Well, really, the human being comes later." Let us show no respect for the human being in embryo, not now. Postpone respect until later—not for this class of human beings.

The role of *desire* is palpable. Researchers *want* this research. The political class *wants* this research. People are making themselves believe, without evidence, and despite many warnings signs to the contrary, that there *will* be glitteringly good results from these experiments, and *only* good results. Desire is getting miles ahead of cool judgment. The extremely plausible horrors of the future are being systematically kept out of the imagination.

I have enormous sympathy for Christopher Reeve, and profoundly hope that, by some miracle of medicine or grace, he is suddenly healed. Yet I am also disappointed by how earnestly one whom I still think of as Superman wants to have other human beings killed, so that he might be cured. He has also plaintively mentioned that some who now don't want embryos destroyed never said a word on their behalf earlier. But in fact, some fifteen years ago, the pope and Cardinal Ratzinger took a lot of heat for their principled opposition to *in vitro* fertilization—that is, to the very techniques of "clumps of cells in a petri dish" and "cells in refrigerators," followed by the wanton destruction of faulty or unwanted embryos, that so many now deplore as less than human. Disagree with their moral conclusions, then or now, if your own mind leads you to. But do not say that they did not grasp, earlier than most, the intellectual principles that are now unfolding before our sentient eyes, and our recoiling imaginations, and our resistant sentiments. Do not say that everyone was silent.

We are testing a great political principle: whether a nation of the people, by the people, and for the people can form great public decisions through open public argument, reflection, and considered choice—or must forever form them by passion and bias and desire and emotional herding.

The president, though he stumbled in his moral reasoning, conceived and executed a shrewd enough political stroke to have temporarily disarmed his foes, won some time, and earned sufficient public standing to lead the nation through a great new era in our history. There was once, in the eighteenth century, a "new science of politics" and , later, a "new science of economics," to both of which our Founders contributed their share of innovations. We are now engaged in learning a "new science of morals," or more exactly a new science of public moral management.

This is the worthiest of tasks for a free society, because what is public freedom for, if not for well-argued and wise moral action? What is the point of political liberty and economic liberty, if we are to live like less-than-human animals?

11
Research with Human Embryonic Stem Cells

Ethical Considerations

Geron Ethics Advisory Board (Karen Lebacqz [Chair],
Michael M. Mendiola, Ted Peters, Ernlé W. D. Young, and
Laurie Zoloth-Dorfman)

RESEARCH WITH EMBRYONIC STEM CELLS

An Ethics Advisory Board, whose members represent a variety of philosophical and theological traditions with a breadth of experience in health care ethics, was created by Geron Corporation in 1998. The board functions as an independent entity, consulting and giving advice to the corporation on ethical aspects of the work Geron sponsors. Members of the board have no financial interest in Geron Corporation.

The Geron Ethics Advisory Board is unanimous in its judgment that research on hES cells can be conducted ethically. In order for such research to be conducted ethically in the current context, some conditions must pertain. In addition, further public discourse will be needed on a range of ethically complex questions generated by this research. The conditions are:

1. The blastocyst must be treated with the respect appropriate to early human embryonic tissue.
2. Women/couples donating blastocysts produced in the process of *in vitro* fertilization must give full and informed consent for the use of the blastocysts in research and in the development of cell lines from that tissue.
3. The research will not involve any cloning for purposes of human reproduction, any transfer to a uterus, or any creation of chimeras.
4. Acquisition and development of the feeder layer necessary for the growth of hES cell lines *in vitro* must not violate accepted norms for human or animal research.
5. All such research must be done in a context of concern for global justice.
6. All such research should be approved by an independent Ethics Advisory Board in addition to an Institutional Review Board.

On 5 November 1998 Geron Corporation announced that scientists working in collaboration with Geron had succeeded in establishing cell culture lines of human embryonic stem (hES) cells. Because these cells are considered pluripotent (capable of being the precursor to a variety of human cell types) and immortal (sustainable in culture and reproducing themselves indefinitely), they represent a major breakthrough in scientific research, with potential for significant advances in tissue transplantation, pharmaceutical testing, and embryology.

Prior to the November announcement, the Geron Ethics Advisory Board developed "A Statement on Human Embryonic Stem Cells"[1] as a set of guidelines for hES research. [See above.] This essay provides an expansion and elaboration of the particular warrants and moral reasoning for that statement.

The EAB did not offer a *carte blanche* approval of research on hES cells undertaken by Geron or any other entity. The board unanimously affirmed that such research can be undertaken ethically, contingent upon meeting a range of qualifying conditions. The initial work of the board has been to specify such conditions; its continued work will consist of assessing Geron's developing research in light of them. Hence here we also include some preliminary reflections on ethical issues in human embryonic germ cell (hEG) research.

Moreover, the EAB perceived the need for urged continued public discussion of the complex ethical issues emerging form such research. Thus the statement and this companion essay should be seen not only as an initial clarification of the EAB's own position, but also as an effort to contribute to and invite that public discourse. We enumerate some specific questions for further reflection at the end of this essay.

1. "The Blastocyst Must Be Treated with the Respect Appropriate to Early Human Embryonic Tissue."

The creation of hES cells involves isolation of cells from the blastocyst.[2] The blastocyst consists of an outer cellular layer, which would develop as the placenta, and an inner cell mass, which would develop as the body of the fetus. The outer layer is dissolved and the resulting mass of cells is used for research. Thus a central ethical issue is the moral status of the blastocyst.

To raise the matter of "moral status" is to ask, Does a given entity possess the requisite qualities or characteristics that entitle it to moral consideration and concern? "Moral status" thus functions as a threshold idea: entities with moral status should be treated in a manner differently form Entities without that status. The EAB affirms that the blastocyst has moral status and hence should be treated with respect.[3]

What sort of moral status does the blastocyst have? This question has riveted political, religious, and ethical attention, and profound and substantial disagreement is based not only on contending biological interpretations but also on deeply held philosophical and theological considerations. Some have argued for conception as the relevant consideration, others for the development of the "primitive streak" (the precursor to the spinal cord of an individual fetus) as a defining moment, and some for utilizing implantation as the crucial threshold for moral status.[4]

Reviewing the complex literature on this topic, Ted Peters, following Daniel Callahan, distinguishes three basic schools of thought.[5] The *genetic school* locates the beginning of human personhood, and thus claims of moral status and dignity, at the genetic beginning—that is, at conception, at the point where one's individual genome is sct. Here, a criterion for moral status (human genetic heritage) is linked to a particular point in human life (conception). The *developmental school*, while granting that human life begins at conception, bolds equally that human *personhood*—and hence *full* moral status—is a later development. Here, moral status is understood developmentally: as the conceptus develops from blastocyst to fetus and beyond so too does moral status grow (although proponents differ on when exactly the threshold of moral status is reached). The *social consequences school* shares with the developmental school the belief that human personhood is a process and an achievement over time. Advocates deny, however, that personhood is achieved at any particular moment. Rather, "personhood" is a matter of definition rather than biological fact, based on socially constructed norms.

In its work, the National Institutes of Health Human Embryo Research Panel focused less on the time when moral status might be acquired and more on the *criteria* for its determination. The panel noted two broad approaches in the debates: one proposes a single criterion as constitutive of moral personhood, while a second, "pluralistic," approach emphasizes a number of different, interacting criteria. As the panel noted

> Among the qualities considered under a pluralistic approach are those mentioned in single-criterion views: genetic uniqueness, potentiality for development, sentience, brain activity, and degree of cognitive development. Other qualities often mentioned are human form, capacity for survival outside the mother's womb, and degree of relational presence (whether to the mother herself or to others). Although none of these qualities is by itself sufficient to establish personhood, their developing presence in an entity increases its moral status until, at some point, full and equal protectability is required.[6]

The panel proposed the pluralistic approach as the more adequate of the two, with moral status (and hence protectability) understood developmentally, culminating at birth in full and equal personhood.[7]

Drawing upon this wealth of philosophical and theological reflections and situating ourselves relative to it, the EAB affirmed our understanding of moral status as developmental and consonant with the pluralistic approach. This developmental view is in accord with Jewish tradition,[8] with the views of many Roman Catholics[9] (although not the Vatican), and with the majority of Protestant traditions as well as with legal traditions that provide different protections at different stages of fetal development.

We hold that a fundamental principle of respect for human life applies at all states of human development. The developmental view that we affirm does not mean that the principle of respect can be ignored; it means that the principle requires different considerations and entails different obligations at different developmental states. For example, Lisa Sowle Cahill argues that "Few doubt that there exists from conception some form of 'human life' in the literal sense; the crucial question is whether from conception or at any subsequent time during pregnancy that life deserves the same respect and protection due an infant."[10] Once there is evidence of capacity for sensation, for example, respect requires minimization of pain. In this very early embryonic tissue there is no capacity for sensation; thus minimization of pain does not apply. Rather, early embryonic tissue is respected by ensuring that it is used with care only in research that incorporates substantive values such as reduction of human suffering.[11] We believe that the purposes of the hES research—its potential to contribute to fundamental knowledge of human development and to medical and pharmaceutical interventions to alleviate suffering—provide such substantive values.

A second source of cells is human embryonic germ (hEG) cells derived from gamete ridge tissue removed from early fetal tissue following elective abortion. These cells have been cultured using similar but not identical methods as are used for the hES cells, and may have both properties of pluripotency and immortality. However, this research raises ethical questions distinct from those in the use of the early blastocyst. The tissue is taken (much as cadaver organs might be taken) from an aborted fetus within the first eight weeks after conception. At stake in this debate is whether licit use can be made of tissue collected after abortion in a society in which the act of abortion is seen by some as murderous and by others (and by law) as acceptable.

The EAB cannot resolve the contentious abortion debate. We are developing guidelines for hEG research. Preliminary reflections suggest at least the following concerns: First, all agree that the demise of the fetus is not caused by the research procedures. Second, the moral obligation to save life may be a sufficiently strong warrant to justify certain uses of the tissue of the dead and hence to support such research. Third, the tissue of the dead must be used with respect. Respect for tissue taken from a dead fetus would take into account the need for closure, grief, and ritual that families might have in these cases. Respect would include the confidential and dignified handling of the tissue when collected and used. Finally, issues of informed consent would apply with the same stringency required in the case of hES research (see below).

2. "Women/Couples Donating Blastocysts Produced in the Process of *In Vitro* Fertilization Must Give Full and Informed Consent for the Use of the Blastocysts in Research and in the Development of Cell Lines from That Tissue."

Human embryonic stem cells are derived from embryos produced from clinical purposes and then donated for research purposes. Hence of crucial ethical significance is the character and quality of the consent given by women and couples to such donation. As with the issue of moral status, the ethical and legal literature on consent and refusal is voluminous.[12] What is needed for valid consent to donate embryonic tissue for research purposes?

Tom Beauchamp and James Childress argue that the core meaning of informed consent is an *autonomous authorization* of a medical intervention or involvement in research. An informed consent occurs "if and only if a patient or subject, with substantial understanding and in substantial absence of control by others, intentionally authorizes a professional to do something."[13] Thus informed consent or refusal is not simply a matter of what the professional should disclose to the patient, but is a matter of ensuring that the conditions necessary for autonomous authorization are met—i.e., that consent is intentional, with substantial understanding, and without controlling influences. Authority for participation in research, expressed in an informed consent, resides with and is to be exercised by the subject.

Consent to utilize embryos for *clinical* purposes if IVF does not therefore suffice as consent for their use in research and cell-line development. Explicit consent must be elicited for such use. We concur with Arthur Caplan that "when research is the goal, whether for profit or not, those whose materials are to be used have a right to know and consent to such use."[14] Other commentators agree that fairness and respect for persons dictate that explicit information be provided to patients regarding the use of their tissues and cells, especially when those may be used for commercial purposes.[15] For these reasons, the EAB determined that women/couples donating embryos for this research should understand clearly the nature of the research and also should understand whether there are commercial implications and if so, whether they hold any proprietary rights in the tissue lines developed from embryonic cells.[16]

Moreover, we believe that the context of embryo donation within the process of IVF renders the need for careful consent even more important. The IVF process is often physically painful, emotionally burdensome, and financially costly. These factors may make IVF patients particularly vulnerable.[17] Such possible vulnerability demands careful and consistent efforts on the part of researchers not to exercise a controlling or coercive influence over these patients. A recent study of patients who consented to participate in therapeutic research found that "physicians' recommendations were . . . powerful factors influencing patients' decisions to become research subjects."[18] Authors of the study recommended that research ethics "be enriched with a sensitivity to the profound trust participants place in researchers and the research enterprise."[19] The context of a study regarding participation in therapeutic research is clearly different from the context of IVF embryo donations for hES cell research. In the latter context, the consent is "twice removed": women/couples are not consenting to therapeutic research, nor does the research involve their own bodies or persons. Nevertheless, even in the IVF context some measure of trust in the researchers may be involved and this possibility raises a warning regarding the ways in which trust can be manipulated to researchers' advantage, thus possibly leading to undue influence over subjects. Moreover, while this research does not involve their own bodies, it does involve entities that they may well consider as potential progeny, a factor that must be considered in this research.

Informed consent also requires "substantial understanding." Hence the EAB requires certain informational elements in the context of embryo donation for hES cell research. First, as the derived hES cells are of significant market potential, donors should understand those potential market implications. Second, they should be advised as to whether there are any proprietary rights in the tissue. The achievement of substantial understanding requires a process that builds trust over time, an admittedly difficult requirement yet one that will be critical in all stages of this research for it to be carried out ethically.[20] Adequate time for eliciting and answering donor questions is necessary, as is ensuring that donors are genuinely free to refuse as well as to consent.

In the course of our deliberations, we reviewed an exemplary form used in soliciting consent for embryo donation for research.[21] This form states explicitly: "cells that may be derived from embryos donated for this research could have clinical and commercial value in the event that the study is successful." It further specifies:

> This research will not benefit your clinical care, and you will not benefit financially from it. The physicians/clinicians involved in your care will not benefit financially from this study. The investigators conducting this study could benefit financially from clinical or commercial values that may result from it. The cells derived in this study may be shared with Geron Corporation, located in Menlo Park, CA, as part of the study. Geron Corporation may benefit financially from the development and clinical use of the cells derived in this study.

Such explicit statements embody well the kinds of financial disclosure that a consent form should contain.

3. "The hES Research Will Not Involve Any Cloning for Purposes of Human Reproduction, Any Transfer to a Uterus, or Any Creation of Chimeras or Human-Animal Hybrids."

Corporate representatives have stated clearly to the EAB that human reproduction was not a goal, purpose, or intent of Geron's research on hES (and hEG) cells. Any effort to produce a living being out of this research would raise a host of ethical issues, in particular the risk of harms to potential offspring,[22] to parenting women and couples, and to the human community itself through genetic manipulation and transmission. Because Geron is not engaged in these activities, and has made clear its intention not to do so, the EAB did not take them up for extended assessment in its deliberations.

Should Geron consider initiating any such activities, the EAB will undertake the necessary ethical analysis.

4. "Acquisition and Development of the Feeder Layer Necessary for the Growth of hES Cell Lines *In Vitro* Must Not Violate Accepted Norms for Human or Animal Research."

To keep hES cells in an undifferentiated state, they are cultured and maintained on layers of nutrients called "feeder layers." Currently, irradiated mouse embryonic fibroblast feeder layers are utilized, but it is possible that other tissues would be use in the future. The acquisition and development of the feeder layers must be in accord with norms for research appropriate to the source from which the feeder layers are drawn.[23]

Geron's use of mouse embryonic cells would seem to fall well within a judgment of ethical use of animal tissue. The research holds great potential therapeutic value. In addition, mouse embryos fall below the threshold of sentience (the ability to experience pain) or of other capacities of organic animal being and activity. Nevertheless, we mandate continued attention to animal welfare.

5. "All Such Research Must Be Done in a Context of Concern for Global Justice."

One of the primary justifications of hES research is beneficence based: its therapeutic potential to alleviate human suffering and to promote the health and well-being of human populations. However, to justify a practice on the basis of its benefits makes moral sense only if people in need actually have access to those benefits. Hence the justification gains credibility only when it is wedded to a commitment to justice, rooted in "a recognition of our sociality and mutual vulnerability to disease and death."[24] *The EAB considers concerns about social justice in public health to be of overriding importance.* Thus in the EAB's judgment, it is morally paramount that research development include attention to the global distribution of and access to these therapeutic interventions.

Two features of Geron's research render this commitment to just access particularly challenging. First, the research is undertaken in the private sector—in the context of market forces, patenting of products, interests of shareholders and investors, and a consideration of profit. These varied interests may compete with—but should not override—a concern for equitable access.[25] Second, the research is highly technological and expensive, as well as under the proprietary rights of a U.S. company. How to ensure adequate access for insured, underinsured, and noninsured patients in the United States, let alone on a global basis, will be an ethically and financially challenging task. The EAB will continue to work with Geron on these matters.

6. "All Such Research Should Be Approved by an Independent Ethics Advisory Board in Addition to an Institutional Review Board."

At its own initiative, Geron established the EAB. Given the kinds of ethical issues already emerging from hES research and those yet to come, the EAB reaffirmed the necessity of an independent board for ethical analysis and consultation. In addition, the research undertaken should receive IRB review and approval in order to protect subjects involved.

ISSUES REQUIRING FURTHER DELIBERATION

As indicated above, the EAB wishes to generate public discourse on a wide range of issues related to emerging research arenas. Among those are some that surfaced during our discussion but are not reviewed in this essay.

1. Who should exercise control over the disposition of fetal or embryo tissue? In our discussion of consent, we refer to both "women" and "couples." The appropriate locus of consent/refusal may be disputed.

2. What is the proper relationship of ethicists to proprietary companies? Who should constitute an ethics board, and who should serve on it? Under what conditions (e.g., remuneration, stock options, etc.)?[26]

3. As noted above, how can difficult issues of global justice and fair distribution be handled in research involving private enterprise?

4. What is the role of consensus in a society that is both plural-

istic and often deeply divided over appropriate norms? How can we develop appropriate language for public debate and decision making while remaining respectful of differences and accountable to substantive moral disagreements?[27]

NOTES

1. The statement was drafted in September 1998 by Karen Lebacqz, revised by EAB, and was finalized on 20 October 1998.

2. The blytocyst is approximately 140 cells, fourteen days postconception.

3. Michael M. Mendiola, "Some Background Thoughts on the Concept of 'Moral Status' Relative to the Early Embryo" (Background paper for the EAB).

4. Jacques Cohen and Robert Lee Hotz, "Toward Policies Regarding Assisted Reproductive Technologies," in *Setting Allocation Priorities: Genetic and Reproductive Technologies*, ed. Robert H. Blank and Andrea Bonnicksen (New York: Columbia University Press, 1992), pp. 228–29.

5. Ted Peters, *For the Love of Children* (Louisville, Ky.: Westminster/John Knox Press, 1996), pp. 96–100.

6. National Institutes of Health, *Report of the Human Embryo Research Panel* (Bethesda, Md.: National Institutes of Health, 1994), p. 49. Mary Anne Warren offers a critique similar in structure, arguing for a "multi-criterial" view of moral status, instead of "unicriterial" approaches. Mary Anne Warren, *Moral Status: Obligations to Persons and Other Living Things* (Oxford: Clarendon Press, 1997).

7. National Institues of Health, *Report of the Human Embryo Research Pannel*, pp. 49–51. Cohen and Holtz echo this position: "New life does not appear suddenly; it is created gradually, with each new phase differing from that which preceded it." Cohen and Holtz, "Toward Policies Regarding Assisted Reproductive Technologies," p. 230.

8. Laurie Zoloth-Dorfman, "The Ethics of the Eighth Day: Jewish Bioethics and Genetic Medicine" (Background paper for EAB).

9. See, for example, Thomas A. Shannon and Allan B. Wolter, "Reflections on the Moral Status of the Pre-Embryo," *Theological Studies* 51 (1990): 603–26.

10. Lisa Sowle Cahill, "Abortion," in *The Westminster Dictionary of Christian Ethics*, ed. James F. Childress and John Macquarrie (Philadelphia: Westminster Press, 1986), p. 3.

11. Ernlé W. D. Young, "The Moral Status of Human Embryonic Tissue" (Background paper for the EAB). The EAB holds that "substantive" values are those supported by prima facie moral duties such as nonmaleficence.

12. See Ruth R. Faden and Tom L. Beauchamp, *A History and Theory of Informed Consent* (New York: Oxford University Press, 1986), for a good overview of the historical development and philosophical treatment of informed consent in the clinical and research contexts.

13. Tom L. Beauchamp and James F. Childress, *Principles of Biomedical Ethics*, 4th ed. (New York: Oxford University Press, 1994), p. 143.

14. Arthur L. Caplan, "Blood, Sweat, Tears, and Profit: The Ethics of the Sale and Use of Patient Derived Materials in Biomedicine," *Clinical Research* 33, no. 4 (1985): 451.

15. See, for example, George J. Annas, "Outrageous Fortune: Selling Other People's Cells," *Hastings Center Report* 20, no. 6 (1990): 36–39; also George J. Annas, "Whose Waste Is It Anyway? The Case of John Moore," *Hastings Center Report* 18, no. 5 (1988): 37–39.

16. The most celebrated legal case dealing with the issue of use of tissue without consent is that of *Moore v. Regents of the University of California* (51 Cal 3rd 120, 1990). For an extended discussion, see E. Richard Gold, *Body Parts: Property Rights and the Ownership of Human Biological Materials* (Washington, D.C.: Georgetown University Press, 1996).

17. Judith Lorber has argued that IVF may also involve a coercive element for women when undertaken to treat male infertility. Such women may have to strike a "patriarchal bargain" to maintain a relationship and have a child within the constraints of monogamy, the nuclear family, and the valorization of biological parenthood. Judith Lorber, "Choice, Gift, or Patriarchal Bargain? Women's Consent to In Vitro Fertilization in Male Fertility," in *Feminist Perspectives in Medical Ethics*, ed. Helen Bequaert Holmes and Laura M. Purdy (Bloomington: Indiana University Press, 1992), pp. 169–80.

18. Nancy E. Kass et al., "Trust: The Fragile Foundation of Contemporary Biomedical Research," *Hastings Center Report* 26, no. 56 (1996): 26.

19. Ibid., p. 27.

20. Mark G. Kuczewiski, "Reconceiving the Family: The Process of Consent in Medical Decisionmaking," *Hastings Center Report* 26, no. 2 (1996): 32.

21. The form was developed by Roger Pedersen, Ph.D., of the Department of Obstetrics, Gynecology, and Reproductive Sciences, University of California, San Francisco, and is used with permission.

22. It is for this reason that the NIH Human Embryo Research Panel distinguished between research on embryos intended for and not intended for transfer. Research on the former must include consideration of potential harms to the future child—a concern not raised by the latter. (National Institutes of Health, *Report of the Human Embryo Research Panel*, pp. 51–52.) It should also be noted that hES cells are derived cells and are not the cellular equivalent of an intact embryo. Even if transferred, hES cells would

not form an embryo because they lack other cells necessary for implantation and embryogenesis.

23. The current usage of mouse embryo cells raises ethical concerns about the use of animals in hES research. A number of criteria have been proposed to weigh ethical versus unethical uses of animals in biomedical research. However, as Strachan Donnelley notes, no single, unambiguous standard or guideline exists for assessing each and every use of animals in science. Ethical attention must be directed to the specific purposes, types, and contexts of animal use. See Strachan Donnelley and Kathleen Nolan, eds. "Animals, Science, and Ethics" [Special Supplement], *Hastings Center Report* 20, no. 3 (1990): S11–S12.

24. Larry R. Churchill, *Rationing Health Care in America: Perceptions and Principles of Justice* (Notre Dame, Ind.: University of Notre Dame Press, 1987), p. 135.

25. Zoloth-Dorfman, "The Ethics of the Eighth Day."

26. Members of the EAB receive a very modest honorarium for the time spent in scheduled meetings with Geron staff and officers. We are not compensated for time spent in research, writing, or other conversations with Geron, nor does any of us hold stock in the corporation.

27. See, for example, Alta Charo, "The Hunting of the Snark: The Moral Status of Embryos, Right to Lifers, and Third World Women," *Stanford Law and Policy Review* 6, no. 2 (1995): 11–38.

12
Ethics and Policy in Embryonic Stem Cell Research

John A. Robertson

M ammalian tissue and organs derive from pluripotent embryonic stem (ES) cells present at the blastocyst stage of embryo development. Embryonic stem cells were first isolated from the inner cell masses of mouse blastocysts in the early 1980s. In November 1998, a team of researchers at the University of Wisconsin and a team at Johns Hopkins University published reports of the first successful isolation and culturing of human ES cells.[1] Because ES cells are capable of self-renewal and differentiation into a wide variety of cell types, the ability to grow them in renewable tissue cultures could have broad applications in research and transplantation.

The research uses of human ES cells include *in vitro* studies of normal human embryogenesis, abnormal development, human gene discovery, and drug and teratogen testing. Potential clinical applications are as a renewable source of cells for tissue transplanta-

John A. Robertson, "Ethics and Policy in Embryonic Stem Cell Research," *Kennedy Institute of Ethics Journal* 9, no. 2 (1999): 109–36. © Johns Hopkins University Press. Reprinted by permission of Johns Hopkins University Press.

tion, cell replacement, and gene therapy. For example, if human ES cells could be directed to differentiate into particular tissues and immunologically altered to prevent rejection after engraftment, they could treat or cure thousands of patients who now suffer from diabetes, neurodegenerative disorders, heart disease, and other illnesses.

The growth of human ES cells in culture is a first but necessary step toward development of cell replacement or regeneration therapies. Future work will have to determine how to obtain human ES cells efficiently and reliably from the inner cell mass of human embryos or from primordial germ cells, grow them in culture, and then identify the growth factors that will direct them to differentiate into cells of particular types to produce the large enough number of pure cells that will be necessary for transplantation.[2] Finally, clinical research using ES-derived cells will be needed to determine under what conditions they are therapeutic for the many conditions that they potentially could treat. Of major importance will be tailoring stem cells genetically to avoid attack by a patient's immune system.

The pace of stem cell research is likely to be affected by ethical and legal concerns beyond the issues of animal data, safety, informed consent, and IRB review that arise with any kind of clinical or human subjects research. The ability to regenerate vital tissues through ES cell technology could eventually transform our understanding of aging and personal identity, as well as place heavy burdens on medical care and social security systems from new ways to extend adult life. Some persons may have serious moral qualms about ES cell research for these reasons alone, and question whether public research support for cell regeneration technologies should be provided at all.

More immediately, ethical and legal issues arise with ES cell research because ES cells have to be derived from aborted fetuses or from the inner cell mass of human preimplantation embryos. Although the ethical and legal issues involved in the retrieval of cells from aborted fetuses and embryos differ in some respects, both sources of ES cells raise concerns about respect for human life at its earliest stages and the extent to which such life may be used or destroyed to provide cells or tissue for research or therapy.

Such questions are highly controversial. Unless ES cells could be obtained from spontaneously aborted fetuses or from preimplantation embryos without destroying them, which appears to be highly unlikely, questions of ES cell research and therapy are likely to rekindle bitter controversies over fetal tissue transplantation and embryo research. Research with more differentiated stem cells, such as mesenchymal stem cells, which already can be directed *in vitro* to become cartilage, bone, fat, tendon, or stroma, will raise fewer ethical and legal problems because those cells can be obtained from consenting persons.[3] However, those cells will not meet the need for many kinds of tissue that can be derived only from pluripotent stem cells. A conflict between respect for the first stages of human life and the persons who will benefit from such research appears to be inevitable.

Legally, research or therapy with fetal tissue derived from induced abortions is now permissible with federal funds. Current federal law, however, prohibits the use of federal funds for the "creation of a human embryo" for research purposes or for "research in which a human embryo or embryos are destroyed, discarded or knowingly subjected to the risk of injury or death."[4] This law broadly defines the term "embryo" to include those that are "derived by fertilization, parthenogenesis, cloning, or any other means from one or more human gametes or diploid cells."

The federal prohibition on the funding of embryo research does not bar ES cell research with private funds in most states, but it slows the progress of such research and creates practical barriers for scientists working in institutions that receive federal research funds.[5] It also leaves researchers without a clear set of ethical or regulatory guidelines for ES cell research with human embryos. With ES cell research now ripe for further advances, and the National Bioethics Advisory Commission (NBAC) and other policymaking bodies considering the issue, it is time to assess the ethical limits on the use of aborted fetuses or IVF-created embryos to obtain ES cells for research and therapy.

ARE HUMAN EMBRYONIC STEM CELLS EMBRYOS?

Ethical questions arise about ES cell research because ES cells must first be derived from human aborted fetuses or preimplantation embryos. It should be clear, however, that ES cells are not themselves embryos because, although they are *pluripotent* in that they could develop into any cell or tissue of the body, they are not *totipotent*. They are not capable of forming a new individual, as a fertilized egg or single cell taken from a four-cell embryo might if cultured *in vitro* and placed in a uterus. Culture of ES cells followed by placement in the uterus would not result in the implantation of an embryo and eventual birth of a child.

General Counsel Harriet Raab of the Department of Health and Human Services relied on this distinction in a legal opinion to the director of the National Institutes of Health on the acceptability under current federal law of research with ES cells. She concluded that the "statutory prohibition on the use of funds appropriated to HHS for human embryo research would not apply to research utilizing human pluripotent stem cells because such cells are not a human embryo within the statutory definition."[6] That definition refers to embryos as "any organism, not protected as a human subject under 45 CFR 46."[7] Referring to a standard dictionary meaning of organism as "an individual constituted to carry out all life functions," Counsel Raab stated:

> Pluripotent stem cells are not organisms and do not have the capacity to develop into an organism that could perform all the life functions of a human being—in this sense they are not even precursors to human organisms. They are, rather, human cells that have the potential to evolve into different types of cells such as blood cells or insulin producing cells.[8]

Counsel Raab's conclusion that ES cell research per se is not covered by the current federal ban on embryo research appears to be legally correct, for ES cells, as she notes, are not themselves embryos nor are they capable of becoming embryos.[9] Her opinion, however, rests on a technical distinction between destructive removal of the ES cells from embryos, which is not permitted with federal funds, and the research that then occurs with those cells after removal. Under Raab's ruling, federal research funding could be provided for research with ES cells as long as the actions involved in removing the ES cells were not federally funded. This technical distinction did not sit well with pro-life members of Congress, who protested that it "would violate both the letter and the spirit of the federal law."[10] Several members of the National Bioethics Advisory Commission also found Raab's conclusion to be "disingenuous" and "legalistic," and objected to authorizing federal support for ES cell research on the basis of that distinction alone.[11] Others might find the distinction useful because it would permit some federal funding of ES cell research without requiring a change in current federal law.[12]

TWO KINDS OF COMPLICITY

Criticism of the HHS Counsel's position rests on the belief that the derivation and use of ES cells cannot be morally or legally separated. If the initial derivation of ES cells from embryos or aborted fetuses is immoral, as many persons believe, then later research uses of those cells do not become morally acceptable merely because the research is separated from the derivation. Under this reasoning, the chain of complicity holds even if the actual ES cells used in research or therapy are the product of many later generations or passages of the original immorally derived ES cells.[13]

Such a view of complicity is a reasonable one up to a point. If the original immoral derivation of ES cells occurred with the intent to make later ES cell research possible, then it is reasonable to view the later researchers as complicit in the original derivation, on the causative theory that the derivation would not have occurred if the later use had not been contemplated. Complicity based on causation, however, would not exist if the original immoral derivation of cells would have occurred regardless of the activities of any particular later researcher. Once the ES cells have been derived for particular kinds of research or for particular researchers, they exist and could be used by other researchers. Under a causative theory of complicity, making the ES cells available to later researchers would not make those researchers complicit in the immoral derivation of the ES cells if their research plans or actions had no effect on whether the original immoral derivation occurred.

The later researchers in this case would, of course, be profiting from or making use of that original immoral derivation and, in that sense, benefitting from another person's wrongdoing. However, a "no benefit from another's wrongdoing" theory of complicity seems much too broad to be a guide to moral or social practice. If taken seriously, it would mean that the taint of an original alleged immoral action, no matter how attenuated, could never be removed as long as it were still traceable to the original action.[14] Such a view would make us all morally complicit in any immoral action that at several removes still underlies or contributes to economic and social transactions from which we benefit. It would also bar the use of organs for transplant that resulted from a homicide or suicide, because the recipient would be benefitting from another's immoral action that occurred independently of the recipient's need for an organ transplant.

ETHICS AND POLICY IN RETRIEVAL OF ES CELLS FROM ABORTED FETUSES

Primordial germ cells retrieved from first-trimester aborted fetuses provide one potential source of ES cells for research and therapy.[15] It is unclear at this time whether fetal tissue or human preimplantation embryos will become the main source of ES cells. Current law, however, permits the use of federal funds to derive such cells from the tissue of aborted fetuses, though it prohibits the use of federal funds to isolate them from embryos.

Because primordial germ cells are removed from fetuses after their death, the derivation of ES cells from aborted fetuses does not cause their death as it does in the case of preimplantation embryos. Nor, with a million-plus electively induced abortions occurring annually in the United States, is there a strong basis for the claim that ES cell research using primordial germ cells would cause many women faced with an unwanted pregnancy to have abortions that would not otherwise have occurred simply because of the chance to donate fetal tissue for research.[16]

Under a causative theory of complicity, neither derivation nor later use of ES cells from abortions that would otherwise have occurred would make one morally complicit in the abortion itself because there is no reasonable basis for thinking that donation of tissue for research after the decision to abort has been made would have caused or brought about the abortion. Thus persons who think that induced abortion is immoral could support the use of fetal tissue or ES cells derived from abortions as long as the derivation or later research or therapy had no reasonable prospect of bringing about abortion, just as they could support organ donation from homicide victims without approving of the homicide that made the organs available. To do so, however, such individuals would have to be convinced that research uses of fetal tissue from abortions otherwise occurring would not bring about future abortions or in some way make abortion appear to be a positive, praiseworthy act.

On the other hand, if persons who think that induced abortion is immoral hold the "no benefit from another's wrongdoing" view of complicity, they would object to the derivation of tissue from fetuses that would be aborted regardless of research plans, merely because one would be deriving benefit from what is viewed as an immoral act. It is unclear how many persons opposed to abortion hold the broader, "no benefit" theory of complicity. If they do hold such a view and are consistent, they should also object to transplanting or receiving organs from murder victims because that would constitute benefitting from another's wrongdoing. Similarly, they should object to any benefits that might in some traceable sense also be the result of another's wrongdoing. Because many common activities, practices, and social arrangements may be traceable to some past wrongdoing—e.g., wresting land from Native Americans—persons holding that view, if they are consistent, would have a difficult time living in the contemporary world.

As a policy matter, the question of obtaining tissue for research or therapy from aborted fetuses received enormous scrutiny between 1988 and 1993 after NIH received proposals to use tissue from aborted fetuses to treat Parkinson's disease, diabetes, and other illnesses. The assistant secretary of health declared a moratorium on NIH funding of fetal tissue transplantation research in 1988 and appointed an expert panel—the NIH Fetal Tissue Transplantation Research Panel—to review the issue. The panel concluded that such research could be ethically conducted if the decision to abort were carefully shielded from the decision to donate tissue, thus reflecting a causative view of moral complicity. In recommending NIH funding for such research, the Panel also recommended that no compensation be provided for donation and that no changes in the timing or method of abortion occur to facilitate retrieval of fetal tissue.[17]

Despite these recommendations, Secretary Louis Sullivan chose in 1990 to continue the federal funding moratorium because he doubted that a strict line between the decisions to abort and to donate tissue could be maintained.[18] This decision was applauded by right-to-life groups but was widely criticized in Congress. Twice Congress passed legislation to overturn the funding moratorium, and twice President Bush vetoed the measures, with the vetoes upheld in each case by a narrow margin. On 22 January 1993, newly elected President Clinton used his executive authority to lift the moratorium on federal funding of fetal tissue transplantation research. Congress then codified authority for federal funding of fetal tissue transplantation research by enacting limits akin to those initially recommended by the NIH advisory panel.[19] That law also made it a federal crime to donate fetal tissue for transplantation to a family member or other designated individual.[20]

Under current federal law it is lawful to conduct and federally fund retrieval of ES cells from primordial germ cells removed from fetal tissue donated after an induced or spontaneous abortion as long as no valuable consideration has been paid for the donation, the woman undergoing the abortion gives informed consent to the dona-tion, and there has been no promise to transplant the tissue into a person specified by the donor or to a relative. In addition, the attending physician must provide a statement concerning the woman's informed consent and the method of obtaining the tissue; a statement by the researcher regarding his or her knowledge of the source of the tissue; and an assurance that the researcher has not participated in any deci-sion regarding termination of the pregnancy. NBAC (1999) has recom-mended that federal funding for retrieval of fetal tissue for stem cell research should continue under these conditions. If these conditions are met, persons who object to abortion but who hold a causative theory of moral complicity have a reasonable basis for finding that retrieval of ES cells from aborted fetuses is ethically and legally accept-able. Since the initial retrieval does not involve causative complicity in the death of the fetus, later research or therapeutic uses of the cells retrieved should be ethically as well as legally acceptable.

ETHICS AND POLICY IN THE RETRIEVAL OF ES CELLS FROM HUMAN PREIMPLANTATION EMBRYOS

A major source of ES cells for research may turn out to be human preimplantation embryos rather than aborted fetuses.[21] As noted above, ES cells derived from embryos are not themselves embryos. However, the need at some point to isolate ES cells for research from live embryos raises ethical concerns of complicity in the destruction of those embryos. Because the embryo is alive at the time that ES cells are removed from the inner cell mass of the blastocyst, research that depends upon the isolation of ES cells from preimplantation embryos raises questions about the permissibility of destroying embryos for research. These questions arise under both the causative and the no-benefit theories of moral complicity, for in either case at least some embryos may be destroyed in order to facilitate or to carry out ES cell research.[22] If the embryos are donated by couples undergoing IVF, the embryos might otherwise have been discarded or kept indefinitely in storage. In some cases they may have been created specifically for the purpose of obtaining ES cells for research. In either case, however, the embryo, unlike the aborted fetus from whom ES cells are retrieved, is alive at the time that the ES cells are removed, even though it may be discarded or never transferred to a uterus if not used in research.

To assess the ethics and policy of ES cell research that is directly destructive of preimplantation embryos, we must revisit debates over embryo research and the status of embryos. Those debates have focused on two main issues: (1) May spare human embryos be destroyed in research when they are donated by couples undergoing IVF treatment for infertility? and (2) May human embryos be created for destructive research without ever having intended to transfer them to the uterus for implantation?

USE OF DONATED SPARE EMBRYOS IN RESEARCH

Normative Concerns

The key ethical and policy debate in embryo research—whether such research should occur at all—has assumed that the embryos used in research would be spare embryos donated by couples undergoing IVF treatment for infertility. Only if such research is deemed acceptable, does the second question—whether embryos may be created for research and then destroyed—arise.

The debate over whether embryos may be destroyed in research when they are donated by couples undergoing IVF has turned on perceptions of the moral status of embryos and the extent to which a rights-based or a deontologic/symbolic view of that status should determine the ethical acceptability of practices involving the first stages of human life. Biologically, there is agreement that the cells of preimplantation human embryos are not yet differentiated into organs or particular tissues. Nor are they clearly committed to individuation because spontaneous twinning can still occur until the time of implantation.[23] Indeed, the embryo's first cellular differentiation is between trophectoderm or placental tissue and the inner cell mass, which contains ES cells. Only after implantation does the embryonic disc and then the primitive streak, from which the brain, the nervous system, and other organs of the body grow, form.

Disagreement, however, attends the moral evaluation of these facts, with differences largely reflecting people's views about abortion. Many people who oppose abortion believe that the fertilized egg's potential to develop into a new human being automatically confers upon it full moral status as a person. They oppose all nontherapeutic embryo research regardless of its benefits, for they view any invasive procedure done to embryos, such as removing stem cells from the inner cell mass, as unjustifiably harming them. Individuals who hold this view are as offended by research with spare IVF embryos as they are with the creation of embryos solely for research. They would oppose any destructive derivation of ES cells from embryos and the research that causes it to occur.

Many other people, however, do not view the previable fetus as an entity that has developed to the point that it has interests that justify overriding a pregnant woman's right to terminate pregnancy. Not surprisingly, persons holding this view about previable fetuses view preimplantation embryos, which are much less developed than fetuses, as too rudimentary in structure or development to have moral status or interests in their own right. For them the attribution of moral status rationally depends upon at least the presence of a nervous system, if not also sentience, and not just its precursor cells. As a result, such individuals maintain that no moral duties are owed to embryos by virtue of their present status and that they are not harmed by research or destruction when no transfer to the uterus is planned. Under this view of embryo status, there is no moral objection to destroying spare, donated embryos to conduct ES cell research. Indeed, if not used in research, those embryos will be discarded or kept indefinitely in frozen storage.

Persons holding the latter view—that the embryo itself lacks interests or rights because of its extremely rudimentary development—do not, however, necessarily view embryos as identical to any other human tissue. Indeed, many such persons would say that embryos, though lacking rights or interests in themselves, deserve "special respect" because of the embryo's potential, if placed in a uterus, to become a fetus and eventually to be born. Even embryos that will not be placed in the uterus have some meaning in this regard for they operate as a symbol of human life or constitute an arena for expressing one's commitment to human life.

This distinction between intrinsic and symbolic valuation of the embryo is at the heart of debate over both abortion and embryo research.[24] One can deny that something has intrinsic value as a moral subject, yet still value it or accord it meaning because of the associations or symbolism that it carries. This distinction has special relevance to the embryo research debate. Even persons who view the embryo as lacking rights or interests in itself are not comfortable with anything at all being done with embryos, for example, using them for toxicology testing of cosmetics or buying and selling them. Accordingly, they would accept the use of spare embryos in research only when there is a good medical or scientific reason for doing so. In effect, the benefits of such research are deemed to outweigh whatever symbolic costs or losses arise from treating an entity that in other circumstances might be transferred to the uterus as if it lacked that potential.

Under this normative approach, embryo research has been deemed acceptable when necessary to pursue a legitimate scientific or medical end that cannot be pursued by other means, when there has been local or national review of the proposed research, and when the embryos have been donated for research with the informed consent of the providing couple. Additional safeguards, such as limiting the research purposes to which embryos might be put or requiring a national ethics advisory board review of such research, would further mark the symbolic importance of spare embryos when they are used in research. The need for additional procedural safeguards depends on the particular political circumstances of the national, state, and institutional context in which they arise. ES stem cell research with spare IVF embryos would clearly meet a high standard of need for conducting research with spare embryos. Its great potential to treat or prevent disease in many persons shows that destructive research with embryos that are unwanted for reproduction will be used for the beneficial purpose of preserving life.

National Advisory Commissions and the Use of Spare Embryos for Research

National advisory commissions in the United Kingdom, Canada, and the United States (as well as in other, although not all, nations) have found embryo research to be ethically acceptable and have recommended government support for it while recognizing the symbolic importance of human embryos, although they have not often specified the normative considerations outlined above. The British acceptance of embryo research was articulated in the landmark 1984 Warnock Report, which found that legally the human preimplantation embryo "did not have the same status as a living child or adult ..." but "... that the preembryo of the human species had a special status."[25] Along with recommendations in many areas for how assisted reproduction should be conducted, the Warnock Report found that research with spare embryos was acceptable for specified purposes. These recommendations laid the foundation for the Human Fertilisation and Embryology Act of 1990, which created a Human Fertilisation and Embryology Authority (HFEA) to license fertility clinics and research projects with human embryos.[26]

The 1990 act, however, prohibits the HFEA from authorizing a research project with embryos "unless it appears to the Authority to be necessary or desirable" to undertake the research for specified purposes. The permissible purposes include treatment of infertility; the acquisition of knowledge about congenital disease, the cause of miscarriage, more effective means of contraception, and preimplantation genetic or chromosomal diagnosis of abnormalities; and "for such other purposes as may be specified by regulations." Under this language, embryos (whether spare or created) could not be used to obtain ES cells to study embryo and tissue development until the Secretary of State for Health issued regulations adding cell and tissue replacement therapies to the list of permitted research purposes.[27]

In the United States, two national ethics review bodies have addressed issues of embryo research prior to the current NBAC attention to the subject.[28] An Ethics Advisory Board (EAB) appointed by the Secretary of Health, Education, and Welfare to advise on federal funding of research involving IVF unanimously agreed in 1979 that "the human preembryo is entitled to profound respect, but this respect does not necessarily encompass the full legal and moral rights attributed to persons."[29] The EAB also found that research with embryos was acceptable for studies related to infertility. Because of then existing DHEW regulations, EAB approval would be required for federally funded embryo research.[30]

When the terms of the EAB members lapsed in the 1980s, neither the Reagan nor Bush administrations appointed new members, thus preventing the requisite EAB approval for federal funding of embryo research. In 1993, the Clinton administration supported a congressional enactment that removed the requirement of EAB review for federally funded embryo research.[31] The director of the National Institutes of Health then appointed a Human Embryo Research Panel (HERP) to recommend guidelines for federal funding of embryo research. The HERP's 125-page report thoroughly canvassed the field and identified many kinds of embryo research, including ES cell research, as acceptable for federal funding with embryos donated by couples undergoing IVF.[32] It also approved the creation of embryos for research when essential to carry out the research. The HERP report clearly rejected the view that embryos have inherent or intrinsic rights, but nevertheless imposed detailed restrictions on their use both to mark their symbolic importance and to show some deference to those who oppose any embryo research. Congress, however, has refused to permit funding of any research with human embryos, regardless of the purpose of the research and source of the embryos.

The promise of ES-based cell or tissue replacement therapies has put the question of federally funded embryo research back on the public agenda, for federal law now prohibits funding of many kinds of ES cell research. At the request of President Clinton, the National Bioethics Advisory Commission has now examined the issue and recommended that "Research involving the derivation and use of ES cells from embryos remaining after infertility treatments should be eligible for federal funding, given an appropriate framework for public oversight and review."[33]

NBAC reached this conclusion after noting that the differences between those who oppose all embryo research and those who support it are largely intractable because "there seems to be no rational way of securing moral agreement in our culture."[34] Instead NBAC sought to develop "policies that demonstrate respect for the alternative points of view and center, where possible, around points of convergence and agreement."[35] It found that there may be a "sufficiently broad, overlapping opinion on the status of embryos to justify, under certain conditions, some research uses of stem cells derived from them."[36] Drawing on Ronald Dworkin's analysis of the acceptance by some abortion opponents of exceptions to a ban on abortion in the case of threat to the mother's life or rape, the NBAC thought that conservatives and liberals might agree on permitting federally funded destructive embryo research "where there is good reason to believe that this destruction is necessary to develop cures for life-threatening or severely debilitating diseases."[37] It also found that the uncertainty about whether such benefits will be forthcoming is outweighed by the great number of potential lives saved if the research leads to successful therapies. Its conclusions and recommendations, however, rest on the assumption that any use of embryos to derive stem cells will occur only when other sources of stem cells for research no longer suffice.

In reaching its conclusion that the ban on federal funding of ES cell research should be partially rescinded, NBAC gave great weight to principles of beneficence and nonmaleficence, which underlay the ethical goals of medicine in conducting research and treating and preventing disease. A ban on federal funding, while permitting private sector research to occur, interferes with such goals and leaves much unregulated embryo research on the fringes of public life. In NBAC's view, "the intentional withholding of federal funds for research that may lead to promising treatments may be considered unjust of unfair."[38]

The overlapping consensus among proponents and some opponents of embryo research described by NBAC assumes limits and regulation in carrying out ES cell research with embryos. Regulatory limits are a key part of that consensus, for they provide accountability, allay public anxiety, and "can ensure greater public accountability and transparency."[39]

To emphasize the sensitivity of such research, NBAC also recommended a national review mechanism for ES cell and other embryo research protocols to "ensure strict adherence to guidelines and standards across the country" and to "provide the public with the assurance that research involving stem cells is being undertaken appropriately."[40] To do this, it recommended establishment of a National Review and Oversight Panel within the Department of Health and Human Services. The Panel would provide "general review of all research protocols in which the use or derivation of human ES cells . . . are considered."[41] Together with a Public Registry of stem cell research proposals, such national ethics review would mark the special importance and meanings attendant on ES cell research and permit public oversight even as it permits such research to occur.

CREATION OF HUMAN EMBRYOS FOR RESEARCH OR THERAPY

Although much ES cell research may occur with cells derived from embryos donated after IVF treatment for infertility, at some point in the future some types of stem cell research, such as efforts to develop or isolate tissue that is compatible with the immune system of prospective patients, may require the creation of embryos. For example, the ability to develop a large library of ES cell genotypes that encode different transplantation antigens so that cell replacement therapies would be available for a large range of the population, likely would depend upon the creation of embryos expressly for that purpose. Or embryos might have to be created by nuclear transfer cloning from a prospective patient's own cells in order to obtain histocompatible ES cells for that patient. The ultimate permissibility of creating embryos solely for ES cell research may thus determine whether many important kinds of ES cell research occur.

The question of the acceptability of the creation of embryos for research has been a heated issue in the embryo research debate, but not between those who oppose and approve of research with spare embryos so much as within the group of those who find destructive research with spare embryos to be morally acceptable. Of course, persons who oppose research with spare embryos on the ground that embryos themselves have intrinsic moral status also oppose the creation of embryos for research. However, persons who approve of research with spare embryos because of their view that embryos lack interests nevertheless disagree about whether embryos should be created for research purposes when there is never any intent to transfer those embryos to the uterus. Thus, a subset of persons exists who approve of research with spare embryos, but object to the creation of embryos solely for research. Presumably these same persons would object to the creation of embryos solely to obtain ES cells for research or therapy. As a result, the ethical controversy over creating embryos specifically for research is likely to surface again if science progresses to the point that creation of embryos becomes necessary to obtain ES cells for certain types of research or therapy.

The opposition to creation of research embryos by persons who approve of research with spare IVF embryos cannot be justified on harm- or rights-based grounds, for those persons also agree that embryos are too rudimentary in development to have interests or rights, and that they are not harmed by destructive research when donated by couples undergoing IVF. Such individuals oppose the creation of research embryos either because of consequentialist concerns about the effect of such practices on other persons or because of deontologic or symbolic/constitutive concerns about showing respect for human life.

The 1994 HERP report mentioned both types of concerns:

> Invoking deeply held and widely shared beliefs about the significance of fertilization as the first step into bringing a potential human being into existence, those opposed to fertilization of oocytes for research argue that this step ought not to be taken solely for research purposes, no matter how important these purposes might be. They maintain that development of embryos expressly for research is *inherently disrespectful of human life*, as well as being open to significant abuses. They also fear that this practice will lead to the *instrumentalization of the preimplantation embryo, and by extension, of other human subjects*. They are particularly concerned that the development of embryos for research *may result in the commodification of embryos and even their commercialization.*[42]

Professors George Annas, Arthur Caplan, and Sherman Elias also have made arguments that combine consequentialist and symbolic deontologic concerns:

To create embryos solely for research—or to sell them, or to use them in toxicity testing—seems morally wrong because *it seems to cheapen the act of procreation* and turns embryos into *commodities.* Creating embryos specifically for research also *puts women at risk* as sources of ova for projects that *provide them no benefit.*

The moral problem with making embryos for research is that as a society we do not want to see *embryos treated as products or as mere objects,* for fear that we will *cheapen the value of parenting, risk commercializing procreation, and trivialize the act of procreation.* It is society's moral attitude toward procreation and the interests of those whose gametes are involved in making the embryos that provide the moral force behind the restriction or prohibition of the manufacture of embryos for nonprocreative uses."[43]

Consequentialist Concerns in Creating Research Embryos

The consequentialist arguments against creating research embryos expressed in these excerpts assume that creation of embryos for well-justified research projects will inevitably lead to bad consequences, yet they never show that those consequences are very likely to occur nor the mechanism by which they would come about. For example, Annas, Caplan, and Elias fear that creating embryos for responsible medical or scientific research will quickly lead to the production of human embryos for more trivial uses such as toxicology screening of drugs or cosmetics and the emergence of an industry that buys and sells human embryos. The HERP quotation cites persons who think that creating embryos will "cheapen or demean" respect for other research subjects. Annas, Caplan, and Elias also think that creating embryos for research will cheapen or demean procreation, thereby undermining respect for persons generally. It will also put at risk women who are induced to serve as donors of the eggs from which the embryos will be created.

It is highly unlikely, however, that such effects would occur if embryos retain the special symbolic respect that marks the HERP and NBAC approaches to research with spare embryos. Given the controversial and sensitive nature of creating embryos for research, it is likely, at least in federally funded research institutions, that research embryos will be created only for compelling reasons, for example, when important research cannot be validly conducted with spare embryos. Creation of research embryos for meritorious research is thus unlikely to lead quickly to a wide use of created embryos for less compelling purposes, such as mass toxicology screening. Nor is a market that buys and sells embryos for research or procreative purposes likely to arise. Indeed, leading proponents of creating embryos for research agree that restrictions on the sale of embryos are desirable.[44]

It is similarly unlikely that the creation of embryos for research that could not easily be conducted in other ways will undermine respect for other research subjects. Apparently the theory for such an effect is that by creating human embryos to serve as instruments or means to obtain knowledge, researchers will be more likely to use other human subjects as mere means. This speculation rests on the counterintuitive premise that carefully limited laboratory research involving the creation of research embryos will affect the interactions that *different* clinical researchers have with human subjects in other projects. The claim also ignores the well-entrenched system of IRB review and informed consent that protects the rights and welfare of research subjects. It is highly implausible that acceptable research projects involving created embryos would undermine or weaken this system or decrease respect for human life or persons generally.

The Annas-Caplan-Elias version of the consequentialist argument—namely, that the creation of research embryos will cheapen or demean human reproduction and parenting by commercializing procreation—assumes an unlikely scenario of embryos being created en masse for commercial purposes. Even if such a practice occurred, it still would not follow that all other acts of conception or procreation would be "demeaned," any more than the existence of prostitution (whether legal or illegal) demeans sexual intercourse between spouses or lovers.[45] In any event, it is difficult to see how the creation of embryos for research when there is a good scientific or clinical need for the practice would "cheapen" or "demean" procreation and parenting generally, much less decrease societal respect for human life, nor do the authors specify how such an effect would come about.

A more substantial consequentialist concern is the effect that the practice of embryo creation might have on women who donate the eggs that are fertilized to create embryos for research.[46] Most of the oocytes that would be fertilized to provide embryos for research are likely to be donated by women who are donating eggs for reproductive purposes to an infertile couple or by women who undergoing IVF but who produce more oocytes than they need for themselves. In some cases, however, researchers may have to recruit women to undergo ovarian hyperstimulation and egg retrieval in order to get eggs for research. The Annas-Caplan-Elias claim that such women would be undergoing significant risks "for no benefit" overlooks the personal benefits those women receive from choosing to serve as contributors to scientific research (a parallel to the altruistic component of oocyte donation for reproductive purposes) and the scientific benefits that the research aims to generate. In any event, all would agree that the rights and welfare of women asked to donate oocytes—whether for reproduction or research—should be carefully protected. They should be fully informed of the risks and benefits of such an act. If they are so informed and nevertheless wish to proceed, there is no justification for permitting egg donation for assisted reproduction but prohibiting it in the case of ES cell research. ES cell research needs are as compelling as the needs of infertile couples.

Deontologic and Symbolic/Constitutive Concerns in Creating Research Embryos

Much of the opposition to the creation of research embryos from persons who reject the notion that embryos have intrinsic moral status rests on deontological or symbolic/constitutive grounds.[47] As the Annas-Caplan-Elias and the HERP quotations show, opponents of creating embryos for research have argued that it is inherently wrong to create an embryo with no intent to transfer it to a uterus because such a practice treats the created embryo as a mere means or instrument to others' ends, thereby expressing "inherent disrespect" for human life. They distinguish the IVF practice of fertilizing more oocytes than can safely be placed in the uterus on the ground that the embryos were created for the purpose of procreation and that *ex ante* fertilization each potential embryo had an equal chance with the others to be transferred to the uterus.[48]

Persons holding the view that the creation of research embryos is wrong try to draw support from the Kantian deontologic tradition that it is wrong to treat human beings as mere means or instruments to the ends of others. This deontologic claim, however, assumes that the preimplantation embryo is already a human being or human subject with interests that are harmed by treatment as a mere means. Yet the persons asserting this ground also agree that spare IVF embryos are too rudimentarily developed to be harmed, and thus may ethically be donated for use in research. Therefore, they cannot object to the creation of research embryos on the ground that doing so would violate ethical norms against using persons as mere means. As they concede, embryos created for research are not persons who can be used to their detriment. If they disagree with this conclusion, then they should object to research with spare embryos as well, for they too become means to serve the needs of others once the decision to use them in research is made.[49]

A more useful way to view nonconsequentialist claims against the creation of embryos for research is as symbolic/constitutive assertions of a person or community's attitudes toward the importance of human life.[50] Because preimplantation embryos are the first stage of a new human life, they ordinarily are created for the purpose of bringing such a life into the world. Accordingly, they function as a powerful symbol of human life and provide the occasion for demonstrating or expressing commitment to human life generally, for example, by condemning creation of embryos for research as being "inherently disrespectful" of human life. In taking such a stance, persons define or constitute themselves as highly protective of human life.

This articulation of the issue tracks very closely the earlier discussion about intrinsic versus symbolic concerns in research with spare embryos and points to a similar resolution. Although embryos do not themselves have rights, they are an occasion for expressing or symbolizing one's views about the importance or value of human life, thereby constituting one's moral or national character in the process. People differ, however, over the degree and the intensity of the symbolic associations that attach to non-rights-bearing entities such as embryos. The importance of signifying or constituting a highly protective attitude toward human life by objecting to certain kinds of embryo research is thus more determined by personal or public policy preferences than it is by the obligations of moral duty.

Individuals might thus accept that the embryo has symbolic/constitutive importance, but find that in particular circumstances other actions connected with the protection of human life also have importance.[51] As we have seen, one could reasonably find that the research and clinical benefits from research with spare IVF embryos outweighs the symbolic detriment that might accrue from using discarded spare embryos as a means of producing knowledge. A similar judgment could be made about the creation of embryos for research. If one concluded that the benefits of the research made possible by the creation of embryos outweighed the symbolic or expressive detriments that some persons perceive to flow from such a practice, then one would find the practice to be ethically acceptable. A good example would be research that creates embryos through nuclear transfer cloning to obtain recipient-compatible tissue or cells for transplant.

In making such judgments, one inevitably compares the symbolic or constitutive costs of a particular research practice with its benefits—e.g., does creating research embryos when necessary to conduct important research so diminish respect for human life that those research benefits should be foregone in order to demonstrate respect for human life in this way? Given the highly personal nature of symbolic/constitutive claims, there is no definitive answer to this question. The answer will depend on the nature and purpose of the research, the availability of alternatives, a person's values and commitments, and public perceptions of a practice.

The ethical acceptability of creating embryos for ES cell research will thus turn on the symbolic/constitutive meanings associated with such a practice in light of the benefits that such research will provide. The conclusion will depend on individual perceptions of the importance of the ES cell research project and the harm to respect for human life that creation of embryos for ES cell research is perceived to cause. Some persons would argue that creation of research embryos raises symbolic or expressive harm beyond that which exists with research with spare embryos and which cannot be justified by the uncertain benefits that the research in question might bring. Other persons, however, might reasonably conclude that the additional symbolic harm of creating embryos for ES cell research that could not otherwise validly occur is minimal. To mark the greater symbolic importance of such research, they might limit it to the most important kinds of research and require additional review procedures.

National Advisory Commissions on
Creating Embryos for Research

The United Kingdom distinguishes acceptable embryo research in terms of the purpose of the research, not the source or mode of creation of the embryos used in the research. This position derives from the 1984 Warnock Committee, which by a nine to seven vote found that embryos could be created solely for research.[52] In 1990 Parliament, in enacting legislation implementing the Warnock Committee recommendations, rejected by a large majority a prohibition on the creation of embryos for research. The Human Fertilisation and Embryology Authority has since licensed several projects involving the creation of embryos for research related to improving infertility treatments. However, until regulations are issued to allow ES cell research with embryos, no embryos, whether spare or created for research, may be used for ES cell research in the United Kingdom.

In the United States, although the 1979 Ethics Advisory Board did not address the question of creating embryos for research, the Human Embryo Research Panel thoroughly addressed the issue in 1994. The Panel recommended, by a large majority, federal funding for creating embryos for research when the research could not otherwise validly occur:

> The Panel recognizes . . . that the preimplantation embryo merits respect as a developing form of human life and should be used in research only for the most serious and compelling reasons. There is also a possibility that if researchers have broad permission to develop embryos for research, more embryos may be created than justified. In order to minimize this, the Panel believes that the use of oocytes expressly for research should be allowed only under the following two conditions:
>
> —When the research by its very nature cannot otherwise be validly conducted. Examples of studies that might meet this condition include oocyte maturation or oocyte freezing followed by fertilization and examination for subsequent developmental viability and chromosomal normalcy and investigations into the process of fertilization itself (including the efficacy of new contraceptives).
>
> —When the fertilization of oocytes is necessary for the validity of a study that is potentially of outstanding scientific and therapeutic value.[53]

The HERP also specifically addressed the creation of embryos for ES cell research. Although the Panel deemed such research not to be acceptable for federal funding in 1994, it found that such research warranted further review in the future. In so recommending, the Panel noted the potential utility of ES cells for transplantation and tissue repair, and circumstances in which creation of embryos to provide ES cells for research might be warranted to create a large library of ES cell genotypes to serve the cell or tissue replacement needs of a large range of the population. It thus implied that the presumption against federal funding to create embryos solely for research might be overcome when a clear research need could be demonstrated.

The Panel also noted that future developments might reduce the need to fertilize oocytes solely for research. For example, the ability to alter ES cell genes to control transplantation antigens would obviate the need to create embryos to obtain a range of lineages compatible with many different genotypes. Similarly, development of the ability to "transfer nuclei from differentiated adult cells into enucleated eggs and obtain normal development to the blastocyst stage" would provide a genome identical to that of the adult donor but would not involve the fertilization of male and female gametes.[54]

The HERP's carefully nuanced recommendations for federal funding of meritorious research that required the creation of research embryos did not fare well in the political process. On the day the Panel's recommendations were announced, President Clinton stated that he opposed federal funding of research embryos regardless of the need or purpose of the research.[55] Congress then banned federal funding of all embryo research, regardless of the purpose or source of embryos.[56] That ban has been reenacted every year since then.

NBAC's report on embryonic stem cell research also discusses the creation of research embryos for ES cell research. Its discussion, however, focuses more on the deontologic or symbolic opposition to such a practice than on consequentialist concerns. It approaches the issue in terms of differences between creating embryos for reproduction and creating them for research. The latter concern appeals to "arguments about respecting human dignity by avoiding instrumental use of human embryos: using embryos merely as a means to some other goal treats them without respect or concern."[57] Such a concern, as I have argued, is, for individuals who do not oppose research on spare embryos, clearly symbolic or expressive, for there can be no obligation to treat a thing as a moral end in itself if it never will have the intrinsic characteristics of a moral entity, which is the case with embryos created for ES cell research.

NBAC, however, did not have to take a position on the issue, for it found that:

> At this time, there are no persuasive reasons to provide federal funds for the purpose of making embryos via IVF solely for the generation of ES cells. More research should be done on stem cells derived from aborted fetuses and embryos remaining after infertility treatment to determine the extent of the need of these additional sources of embryos for research.[58]

If the science already had progressed to the point that there were a strong medical or scientific need to create embryos for ES cell research (because spare embryos were not available for donation or because person-specific embryos were needed), NBAC would have conceivably taken a different position. The unarticulated symbolic/expressive normative approach that underlies the NBAC position on ES cell research with spare embryos would, as a similar position did for the HERP, permit approval of the creation of embryos for ES cell research when necessary to carry out important research, such as the study of embryos created by cloning to provide histocompatible tissue for transplant.

PUBLIC RESEARCH POLICY AT THE BEGINNING OF LIFE

With advances now rapidly occurring in the understanding and manipulation of human ES cells, current federal policy against funding any destructive embryo research needs to be reassessed to avoid discouraging or preventing such research. The Raab distinction between federal funding of use rather than derivation of ES cells may have some political appeal[59] because it would permit some federally funded ES cell research to proceed without necessitating a change in the present law, as federal support of ES cell derivation would require. A ban, however, on federal funding of either the derivation or the use of ES cells creates a barrier to further progress in ES technology. For example, permitting federally funded research with ES cells derived in the private sector still leaves researchers facing intellectual property and other barriers to obtaining privately derived ES cells. Those barriers could deter some researchers from undertaking ES cell research and thus slow the pace of progress in the field.

Only the group of persons who view fertilized eggs, early embryos, and fetuses as themselves persons or subjects with intrinsic rights strongly object to the use of aborted fetuses or spare embryos for ES cell research.[60] Their view of the intrinsic moral status of embryos—that spare or created embryos are persons that cannot be destroyed in research—is not required by our legal and constitutional traditions and should not drive federal research policy. Indeed, the basis in this position for imputing intrinsic value to the embryo is more the reflection of religious or spiritual perspectives than it is of a persuasive philosophical or normative position about when protected human life begins. It has been rejected by most advisory commissions that have studied the matter.

Precisely because people differ so deeply over personal spiritual and value commitments, one group should not erect its own view of the matter into public policy. Indeed, doing so exacts a high cost from those who would benefit from such research. Their health needs are held hostage to a set of religious positions held by a minority that has inordinate sway in the legislative process. Prohibition of federal funding for ES cell research is also a major barrier to further research and development in this field. In addition, it leaves private sector researchers without a clear set of ethical or regulatory guidelines for the ES cell research, which they do, and will continue to, conduct.

A more fruitful approach to research issues at the beginning of life is to recognize that for most persons the ethical or normative questions that arise are less about the duties intrinsically owed to embryos or fetuses than they are about symbolizing or expressing the high respect that most persons have for human life generally. This respect is shown, not by banning all research with embryos or aborted fetuses, but by allowing such research only when good reasons exist for engaging in it and an institutional, or even national, review process to assess those reasons has been implemented.

National review bodies in the United Kingdom and the United

States that have considered embryo research have, with some variation in the details of justification, taken such a normative approach to issues of embryo research. They have consistently refused to find that embryos are themselves persons with intrinsic rights, and they have recommended that research be permitted with spare embryos when necessary for good medical or scientific purposes. The Warnock Committee and the NIH Human Embryo Research Panel also have supported the creation of embryos for research when important research could not otherwise occur, including ES cell research when that is shown to be necessary.

NBAC has re-analyzed these questions in the context of ES cell research and has come to similar conclusions. Congress should enact NBAC's recommendations and allow federal support for both the derivation and the use of ES cells from spare embryos to occur under a system of national review. If ES science progresses to the point that researchers have to create embryos with particular genomes in order to overcome immunological incompatibility in prospective patients, then Congress should also provide federal funds for that research.[61] Embryo research policies, when structured and regulated to mark the special importance of this category of research, can reinforce respect for the earliest stages of human life even as material from those stages is used to help others.

NOTES

1. James A. Thomson et al., "Embryonic Stem Cell Lines Derived from Human Blastocysts," *Science* 282 (1998): 1145–47; Michael J. Shamblott et al., "Derivation of Pluripotent Stem Cells from Cultured Human Primordial Germ Cell," *Proceedings of the National Academy of Sciences USA* 95 (1998): 13726–31.

2. Eliot Marshall, "A Versatile Cell Line Raises Scientific Hopes, Legal Questions," *Science* 282 (1998): 1014–15.

3. Gretchen Vogel, "Harnessing the Power of Stem Cells," *Science* 283 (1999): 1432.

4. Public Law 105–277, H.R. 4328, sect. 511 (21 October 1998).

5. Eliot Marshall, "Varmus Grilled over Breach of Embryo Research Ban," *Science* 276 (1997): 1963.

6. Harriet Raab, *Federal Funding for Research Involving Human Pluripotent Stem Cells*, memorandum to Harold Varmus, Director, NIH, January 15, 1999.

7. Ibid., p. 2. Although the term "any organism" could refer to non-human organisms as well, it is clear from the context that the statute in question is meant to apply to human organisms.

8. Ibid., pp. 2–3.

9. One research team has reported that mouse ES cells, when combined with trophectoderm cells from another mouse embryo, could produce a new individual, thereby undercutting the claim that pluripotent ES cells are not totipotent in that they could not produce the birth of another individual. (See Andras et al., "Derivation of Completely Cell Culture–Derived Mice from Early-Passage Embryonic Stem Cells," *Proceedings of the National Academy of Sciences USA*, 90 [1993]: 8424–28.) If this claim is true, it would undercut the biologic basis for HHS Counsel Raab's legal opinion that pluripotent cells are not embryos within the meaning of the current federal ban on federal funding of embryo research. The claim would not, however, prevent the use of such cells in research or therapy if one accepted that totipotent cells or embryos could ethically be used in destructive research. The author thanks Andrea Bonnicksen for helpful comments on an earlier draft and Lee Silver for bringing the study to my attention.

10. Nicholas Wade, "Ruling in Favor of Stem Cell Research Draws Fire of Seventy Lawmakers," *New York Times*, February, 17, A12.

11. Matthew Davis, "Presidential Bioethics Advisors May Come Down on the Side of Federal Funding of Research on Stem Cells Derived from Embryos," *Washington Fax*, March, 4, 1999, 1–2.

12. Matthew Davis, "Administration Supports Publically Funded Stem Cell Research, But Not to the Point of Seeking Changes in the Embryo Research Ban," *Washington Fax*, July 19, 1999, 1–3.

13. Many persons, of course, do not think that the original derivation of ES cells either from aborted fetuses or the abortions that make such donation possible or from live embryos is immoral, and they thus do not see the problem of moral complicity that those who object to ES cell removal do.

14. For example, the United States Supreme Court's equal protection jurisprudence has rejected a "benefit from" view of past racial and gender discrimination in assessing race or gender-neutral public policies that have a disproportionate or disparate racial or gender impact. See *Massachusetts v. Feeny* (442 U.S. 256 [1979]), in which past discrimination against women by the military does not render state preferences for veterans for civil service jobs discriminatory under the fourteenth amendment.

15. Shamblott et al., "Derivation of Pluripotent Stem Cells."

16. National Bioethics Advisory Commission (NBAC), *The Ethical Use of Human Stem Cells in Research*, draft report, July 9, 1999, III–4.

17. National Institutes of Health (NIH), *Report of the Panel on Fetal Tissue*

Transplantation Research (Bethesda, Md.: NIH, 1989). The Polkinghome Committee, which advised the United Kingdom government on fetal tissue transplantation, made similar recommendations. (See U.K., *Review of the Guidance on the Research Use of Fetuses and Fetal Material*, Department of Health and Social Security, 1989).

18. John A. Robertson, *Children of Choice: Freedom and the New Reproductive Technologies* (Princeton, N.J.: Princeton University Press, 1994), p. 209.

19. *NIH Health Revitalization Act of 1993*, Public Law 103–43, sect. 111, 107.

20. John A. Robertson, "Abortion to Obtain Fetal Tissue for Transplant," *Suffolk University Law Review* 27 (1993): 1362–69.

21. Thomson et al., "Embryonic Stem Cell Lines."

22. Persons who hold a causative theory of complicity and who believe that embryo destruction is immoral should object to ES cell research only when it can reasonably be shown that embryos would not have been destroyed if the ES cell research had not been planned or contemplated. However, those same persons need not object to ES cell research that cannot reasonably be shown to have brought about the destruction of the embryos.

23. Clifford Grobstein, "The early development of human embryos," *Journal of Medicine and Philosophy* 10 (1985): 213–20.

24. Robertson, *Children of Choice*. Ronald Dworkin makes a similar distinction in his discussion of abortion in *Life's Dominion* (New York: Alfred A. Knopf, 1993) with his use of the labels "derivative" and "detached" to correspond to the notions of "intrinsic" or "rights-based" and "symbolic," "expressive," or "constitutive" protection used here. John Fletcher ("Deliberately Incrementally on Human Pluripotential Stem Cell Research," NBAC background papers on embryonic stem cell research, 1999), in an important background paper prepared for NBAC, has skillfully shown how Dworkin's distinction helps cross the gap to those who highly value early embryos and fetuses. His overall analysis is similar to the analysis contained in this article.

25. U.K., Report of the Committee of Inquiry into Human Fertilization and Embryology, Department of Health and Social Security, 1984.

26. U.K., Human Fertilization and Embryology Act, 1990.

27. Eliot Marshall, "Britain Urged to Expand Embryo Studies," *Science* 282 (1999): 2167–68.

28. Some state laws, which resulted from state efforts in the mid-1970s to block fetal research in the wake of *Roe v. Wade*'s 1973 constitutional recognition of a right to abortion, make research with human embryos for any purpose and from any source a crime. (See John A. Robertson, "Embryo Research," *Western Ontario Law Review* 24 (1986): 15–37 and Lori Andrews, "State Regulation of Embryo Research" in *Papers Commissioned for the Human Embryo Research Panel* (Bethesda, Md.: National Institutes of

Health, 1994). Federal courts have struck down bans on embryo research in three states (Louisiana, Illinois, and Utah) on grounds of vagueness and interference with procreative rights (*Margaret S. v. Edwards*, 794 F.2d 994 [5th Cir. 1986]; *Lifchez v. Hartigan*, 735 F. Supp. 1361 [N.D. Ill], affd 914 F.2d 260 [7th Cir. 1990], cert. denied sub nom., *Scholberg v. Lifchez*, 498 U.S. 1069 [1991]; *Jane L. v. Bangerter*, 61 F.3d 1493 [10th Cir. 1995]). Laws in the states that now ban embryo research will remain a significant barrier to ES cell research regardless of the source of research funding, the purpose of the research, or the source of embryos. However, those laws would not ban research with ES cells once they have been isolated or cultured, for they would then no longer fit the definition of embryo or fetus contained in those statutes.

29. U.S. Department of Health, Education, and Welfare (DHEW), *HEW Support of Reseach Involving Human In-Vitro Fertilization and Embryo Transfer*, 1979, p. 44; Ethics Advisory Board, "Report and Conclusions," *Federal Register* 44 (1979): 35.033–58.

30. "Research Involving Pregnant Women of Fetuses," *Federal Register* 45 (2001): 46.204(d).

31. Public Law 103–43, 121 (c) (1993).

32. NIH, *Report of the Human Embryo Research Panel*, 1994.

33. NBAC, *The Ethical Use of Human Stem Cells*, p. IV–9.

34. Ibid., p. IV–14, quoting Alastair McIntyre.

35. Ibid., p. IV–16.

36. Ibid.

37. Ibid., p. IV–18.

38. Ibid., p. V–9.

39. Ibid., p. VI–17.

40. Ibid., p. VI–20.

41. Ibid., p. V–22.

42. NIH, *Report of the Human Embryo Research Panel*, p. 42, emphasis added.

43. George Annas, Arthur Caplan, and Sherman Elias, "The Politics of Human-Embryo Research—Avoiding Ethical Gridlock," *New England Journal of Medicine* 334 (1996): 1331.

44. NIH, *Report of the Human Embryo Research Panel*.

45. Margaret Jane Radin, *Contested Commodities* (Cambridge: Harvard University Press, 1996); Scott Altman, "(Com)modifying Experience," *Southern California Law Review* 65 (1991): 293–340.

46. Nicole Gerrand, "Creating Embryos for Research," *Journal of Applied Philosophy* 10 (1993): 175–87.

47. In Dworkin's terminology, such concerns are "detached" from the interests of actual persons and not "derivative" of their status as rights-holders.

48. NIH, *Report of the Human Embryo Research Panel*; Annas, Caplan, and Elias, "The Politics of Human-Embryo Research."

49. Gerrand, "Creating Embryos for Research."

50. John A. Robertson, "Symbolic Issues in Embryo Research," *Hastings Center Report* 25, no. 1 (1995): 37–38.

51. Such a cost-benefit judgment explains the acceptability in IVF therapy of fertilizing more oocytes than can be safely transferred to the uterus, with the excess then discarded or never transferred. Although this practice causes embryos to be created and then destroyed in order to achieve pregnancy and childbirth, it serves a legitimate, valued need that negates the disrespect for human life that symbolically might occur if embryos were created and then destroyed for trivial purposes.

52. U.K., *Report of the Committee of Inquiry into Human Fertilization and Embryology*.

53. NIH, *Report of the Human Embryo Research Panel*, pp. 44–45.

54. Ibid., p. 79.

55. William Jefferson Clinton, "Statement by the President on NIH Recommendation Regarding Human Embryo Research," *U.S. Newswire*, December 2, 1994.

56. Public Law 104–99, sect. 128 (January 26, 1996).

57. NBAC, *The Ethical Use of Stem Cells in Research*, p. V–10.

58. Ibid., p. V–10.

59. Davis, "Administration Supports Publically Funded Stem Cell Research."

60. If such individuals hold a causative view of complicity, they should accept derivation of primordial germ cells from aborted fetuses when the derivation is clearly separated from the decision to abort, as now exists in federal law. They should also accept ES cell research with cells destructively derived from embryos as long as the research was not a cause of the derivation.

61. If creation of embryos is necessary to obtain ES cells of particular genotypes, as might occur to create a universal library of replacement tissue or to tailor replacement tissue to the recipient via nuclear transfer cloning, then fertilization of oocytes or creation of embryos through cloning would be "necessary for the validity of the study," and should be permitted. (See John A. Robertson, "Two Models of Human Cloning," *Hofstra Law Review* 27 [1999]: 609–39.)

13

The Ethics and Politics of Small Sacrifices in Stem Cell Research

Glenn McGee and Arthur Caplan

Pluripotent human stem cell research may offer new treatments for hundreds of diseases,[1] but opponents of such research argue that pluripotent stem cell therapy comes attached to a Faustian bargain: the destruction of many frozen embryos for every new cure. The National Bioethics Advisory Commission (NBAC), the Geron Ethics Advisory Board (GEAB), and many scholars of bioethics, including, in these pages, John Robertson, have raised interesting ethical issues about potential stem cell research.[2] But nothing holds the attention of the public or lawmakers like the brewing battle between millions of ill Americans who favor stem cell research and millions who oppose the destruction of any fetus or embryo for any purpose. The plan to sacrifice embryos for a revolutionary new kind of research has reawakened a long-dormant academic debate about the morality of destroying developing human life.

Glenn McGee and Arthur Caplan, "The Ethics and Politics of Small Sacrifices in Stem Cell Research," *Kennedy Institute of Ethics Journal* 9, no. 2 (1999): 151–58. © Johns Hopkins University Press. Reprinted by permission of Johns Hopkins University Press.

In these pages, Robertson (like the GEAB and NBAC) argues that one's position on whether pluripotent stem cells can be used in research depends on one's view of the intrinsic and symbolic status of the embryo.[3] In essence, Robertson and many others argue that one's position on stem cell research pretty much boils down to one's position on abortion. But that is too simple a framework.

Robertson and NBAC make two mistakes. First, they duck the task of identifying and analyzing criteria for justifying the sacrifice of human cells and human life in the name of research. All who write about pluripotent stem cells refer to respect for pro-life views about symbolic moral status, and GEAB in particular claims that symbolic status is the reason such research should be therapeutic in nature. Yet so far no one has explained what kind of destruction is appropriate and of what sort of creatures and for what specific goals and contexts.

Second, Robertson, NBAC, and GEAB take what we call an "accommodationist" posture toward political opposition to stem cell research. In their haste to avoid rehashing the abortion debate, many proponents of stem cell research, including the National Institutes of Health Counsel, Harriet Raab,[4] cede too much ground to foes of abortion procedures, allowing too many arguments to go unanswered. They thus miss critical opportunities to engage in ethical debate about what sorts of rights and duties attend the making and management of embryonic life.

MORAL SACRIFICE

It seems to us that the central moral issues in stem cell research have less to do with abortion than with the criteria for moral sacrifices of human life. Those who inveigh against the derivation and use of pluripotent stem cells make the assumption that an embryo has not only the moral status of human person, but also a sort of super status that outweighs the needs of others in the human community. It is wrong to abort or kill a human, they argue, and thus it is wrong to kill an embryo. But this argument, which is problematic when made about abortion more generally, is doubly so when made against the derivation of pluripotent stem cells from embryos.

Even if frozen human embryos are persons, symbolically or

intrinsically, this in no way entails the right of a frozen embryo to gestation, or to a risk-free pathway into maturation. Adult and child human beings' right "to life" is, considered constitutionally and as a moral problem, at best a negative right against unwarranted violence by the state or individuals. There are, sadly, few positive rights involved. Americans cannot, for example, claim a right against the state to protect them against disease, disasters, adverse weather, and other acts of nature. If a frozen human embryo is a full human person, it still has no right to life per se, but rather a negative right against unwarranted violence and a weak positive right to a set of basic social services (police protection, fire protection, and the like). The question remains as to what constitutes unwarranted violence against an embryo, and for what reasons might an embryo ethically be destroyed—e.g., in the interest of saving the community. Adults and even children are sometimes forced to give life, but only in the defense or at least interest of the community's highest ideals and most pressing interests. One would expect that the destruction of embryonic life, whatever its moral status, would also take place only under the most scrupulous conditions and for the best communal reasons. It bears noting that only those who consistently oppose all violence, destruction, or killing of any kind in the name of the state, the church, or the community can rationally oppose the destruction of an embryo solely by virtue of its status as a human person.

It remains to be shown what the common good is, and what sort of sacrifice an embryo should make in its interest. It is commonly held that no human being should be allowed to lie unaided in preventable pain and suffering. The desire to ameliorate the suffering of the ill motivated Hippocrates, St. Francis of Assisi, Cicero, and Florence Nightingale. It is a central tenet of contemporary medicine that disease is almost always to be attended to and treated because it brings such pain and suffering to its victims and to their family and communities. Trade-offs are made in the treatment of disease, against cost and other competing social demands. But both the Western ethic of rescue and the practical structure of contemporary health care and other social institutions make it clear that among the deepest moral habits of human life is that of compassion for the sick and vulnerable. One of the compelling tenets of the movement to prevent abortion is the argument that a pregnancy ought not be terminated for superficial reasons, but should be viewed as a responsibility to aid the developing human life and to prevent it from need-less suffering.

It is the moral imperative of compassion that compels stem cell research. Stem cell research consortium Patient's CURe estimates that as many as 128 million Americans suffer from diseases that might respond to pluripotent stem cell therapies. Even if that is an optimistic number, many clinical researchers and cell biologists hold that stem cell therapies will be critical in treating cancer, heart disease, and degenerative diseases of aging such as Parkinson's disease. More than half of the world's population will suffer at some point in life with one of these three conditions, and more humans die every year from cancer than were killed in both the Kosovo and Vietnam conflicts. Stem cell research is a pursuit of known and important moral goods.

WHAT IS DESTROYED?

The sacrifice of frozen embryos is a curious matter. Set aside, again, the question of whether a frozen embryo is a human life or human person. Grant for a moment that a one hundred–cell human blasto-cyst, approximately the size of the tip of an eyelash and totally lacking in cellular differentiation, is a fully human person. What does such a person's identity mean, and in what ways can it be destroyed? What would it mean for such a person to die? When could such a death be justified? These questions require a new kind of analysis.

The human embryo from which stem cells are to be taken is an undifferentiated embryo. It contains mitochondria, cytoplasm, and the DNA of mother and father within an egg wall (which also contains some RNA). None of the identity of that embryo is wrapped up in its memory of its origins: it has no brain cells to think, no muscle cells to exercise, no habits. The one hundred–cell embryo has one interesting and redeeming feature, which as best anyone can tell is the only thing unique about it: its recombined DNA. The DNA of the embryo contains the instructions of germ cells from father and mother, and the earliest moments of its conception determined how the DNA of mother and father would be uniquely combined into a human person. The DNA of that person will, if the embryo survives implantation, gestation, and birth, continue to direct many facets of the growth and identity of our human person. At one hundred cells, nuclear DNA is the only feature of the embryo that is not replaceable by donor components without compromising the critical features of the initial recombination of maternal and paternal genetic material after sex (or, in this case, *in vitro*).

Opponents of stem cell research make an antiabortion argument, namely that the harvesting of pluripotent stem cells will require the destruction of the embryo. But while the cytoplasm, egg wall, and mitochondria of the embryo are destroyed, we just noted that none of these cellular components identifies the embryo at the one hundred–cell stage. The personifying feature of a one hundred–cell blastocyst is its DNA. Pluripotent stem cells from the harvested embryo are directed to form cell lines, each cell of which contains, in dormant form, the full component of embryonic DNA. The DNA in the cell lines has a much greater chance of continuing to exist through many years than does the DNA of a frozen embryo (which in most cases already will have been slated for destruction by the IVF clinic that facilitated the donation, and which would have no better than a 5 to 10 percent chance of successful implantation in any event). Although most Americans are opposed to the "cloning" of adult human beings, it might be possible to harvest DNA from any of the stem cell–based cell lines to make a new, nuclear transfer–derived embryo, or in fact to make five or ten embryos, each of which would possess all of the DNA of the original embryo. In this sense, the critical, identifying features of the embryo would never have been destroyed in the first place, unless what one means by "destroying" an embryo is the loss of its first egg wall, cytoplasm, and mitochondria. The transfer of the nucleus from an embryo to an enucleated egg is a bit like a transplant, though here the donor and the donation are both the DNA. In the case of embryos already slated to be discarded after IVF, the use of stem cells may actually lend permanence to the embryo. Our point here is that the sacrifice of an early embryo, whether it involves a human person or not, is not the same as the sacrifice of an adult because the life of a one hundred–cell embryo is contained in its cells' nuclear material.

The task of balancing sacrifice in the community is one encountered by Solomon in the Judeo-Christian religious texts. Our institutions must enable us in the community to debate and identify the ideals that merit sacrifice, and the loss must be weighed with justice in mind. An embryo cannot reason and it cannot reject a sacrifice or get up and leave the community. For those who feel special responsibility to embryos, the vulnerability of the frozen embryo may suggest special consideration of the kind given to all moral actors in society who are for one reason or another without voice. The question remains, though: What need is so great that it rises to the level where every member of the human family, even the smallest of humans might sacrifice? Already it is clear that we believe that no need is more obvious or compelling than the suffering of half the world at the hand of miserable disease. Not even the most insidious dictator could dream up a chemical war campaign as horrific as the devastation wrought by Parkinson's disease, which destroys our grandparents, parents, and finally many of us.

If Parkinson's disease, or for that matter if a dictator could only be stopped through the destruction of an infant, most every human would blanch at the idea of such sacrifice. But that is not what is asked here. Even those who hold that an embryo is a person will not want to argue that the life of a one hundred–cell embryo is contained in its inessential components. Assuming that a developing embryo can be salvaged by transplanting its DNA, as we have described, it seems unreasonable to oppose the destruction of the embryo's external cellular material or to fear that the one hundred–cell embryo is killed in the transfer. The identity of the human embryo person, if it is a symbolic or intrinsic person, is tied at that stage to the DNA. That DNA is not lost or even injured by the harvesting of embryonic stem cells. This is not the sacrifice of the smallest and most vulnerable among us. We are debating the potential for temporary transplant of undifferentiated tissue and the DNA of such a "person," rather than the imminent destruction of an embryo discarded by a clinic. It is difficult to imagine those who favor just war opposing a war against such suffering given the meager loss of a few cellular components.

POLITICAL ACCOMMODATION

The road to a democratic debate about stem cell research is difficult, as Robertson acknowledges. We entirely agree with Robertson that NIH, and researchers working on stem cell–based cell lines, need not be considered complicit in the destruction of an embryo far removed in time and space.[5] However, complicity assumes someone has done something wrong. Critics and NBAC are thus right to call attention to the amoral nature of the Raab opinion. It could not be more plain that NIH (and some politicians who favor stem cell research) initially hoped to sidestep entirely any ethical debate over the use of cell lines derived from the destruction of embryos. As we have argued above, it hardly does justice to those who believe that embryos are human persons to sacrifice those persons in a cloud of deception and cowardice. Moral sacrifice demands the highest accountability of our social institutions and especially those entrusted with writing and litigating the theory and policy that guides bodies like NIH.

Since at least 1980, the market on "family values" has been cornered by one side of the political spectrum. Anticipating that the public prefers platitudes about abortion to complex positions, most in American political office do not even discuss abortion and embryo research beyond a simple reference to their pro- or anti-abortion voting record. Meanwhile, the last twenty years have seen conservatives develop extraordinarily rigid and arcane positions on virtually every area of "family values."

Again and again, bioethicists' and politicians' policy of appeasing conservative views on family values has resulted in bad policy and balkanized public debate. Embryo research and fetal tissue are particularly instructive in this regard. Without an engagement of the moral questions involved and without an explicit moral framework to explain the decisions, the nation found itself with bans on embryo research and fetal tissue transplantation research that the majority of citizens did not support.

NIH Counsel Raab, National Institutes of Health Director Harold Varmus, NBAC, the GEAB, and in these pages Robertson each give too much moral ground by ceding the illogical (embryos are special people who can never be allowed to die) and bizzare (researchers must not be complicit in the death of an embryo) arguments made by opponents of stem cell research. The sight of scientists rushing to find a legal opinion that legitimizes their work is not a pretty one, nor is that of politicians running scared from those who threaten to shut down the National Institutes of Health in the interest of preserving frozen embryos in stasis. We must be conscious of the role of politics,[6] but the law and the facts do not help with stem cell research. The issues here are novel and they are hard, but mostly they require philosophical innovation about what an embryo is and how we are to treat embryonic material in a time of stem cell research. Our argument here is that no embryo need be sacrificed, but we must alter the terms and goals of our debate to frame an appropriate moral framework for dealing with embryos.

NOTES

1. James A. Thomson et al., "Embryonic Stem Cell Lines Derived from Human Blastocysts," *Science* (1998): 1145–47; Gretchen Vogel, "Harnessing the Power of Stem Cells," *Science* 283 (1999): 1432.

2. National Bioethics Advisory Commission (NBAC), *The Ethical Use of Human Stem Cells in Research*, draft report, July 9, 1999; Geron Ethics Advisory Board (GEAB), "Research with Human Embryonic Stem Cells: Ethical Considerations," *Hastings Center Report* 29, no. 2 (1999): 36–38; Glenn McGee and Arthur Caplan, "What's in the Dish? [Symposium on ethical issues in stem cell research]," *Hastings Center Report* 29, no. 2 (1999): 38–41; John Fletcher, "Deliberately Incrementally on Human Pluripotential Stem Cell Research," NBAC background papers on embryonic stem cell research, 1999; John A. Robertson, "Ethics and Policy in Embryonic Stem Cell Research," *Kennedy Institute of Ethics Journal* 9 (1999): 109–36.

3. Robertson, "Ethics and Policy in Embryonic Stem Cell Research."

4. Harriet Raab, *Federal Funding for Research Involving Human Pluripotent Stem Cells*, memorandum to Harold Varmus, Director, NIH, January 15, 1999.

5. Robertson, "Ethics and Policy in Embryonic Stem Cell Research."

6. Glenn McGee, "Pragmatic Method in Bioethics" in *Pragmatic Bioethics*, ed. Glenn McGee (Nashville, Tenn.: Vanderbilt University Press, 1999), pp. 31–73.

Part 4
Religious Issues

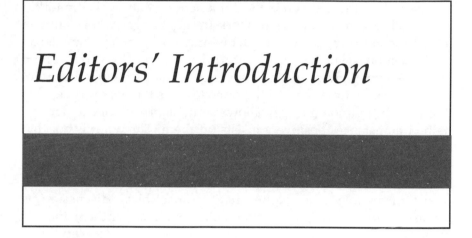

Editors' Introduction

We start this section with two short pieces intended to show just how complicated this whole issue truly is. The first is testimony from a young boy who has suffered dreadfully from disease and yet whose faith demands that we abstain activities that demand the destruction of embryos. One cannot but admire such a commitment. But should—need—the rest of us be ruled by it? The second piece, on the rights and demands of the sick, makes one wonder if one group in society ever has the moral authority to impose their religiously based beliefs on other groups in society. The contributions that make up the rest of this section show that there is no straightforward answer to this query. Religious people are as divided as are the nonreligious. A fairly liberal Protestant like Ronald Cole-Turner finds a place for ethical stem cell research. More conservative churches do not, as is shown by the testimony of Greek Orthodox priest Demetrios Demopulos. Likewise, as Aline Kalbian shows, on both biblical and philosophical grounds the Catholic Church is opposed absolutely to any work that tampers with the integrity of the unborn.

Note that there are no biblical injunctions that explicitly forbid stem cell research. One has to work from passages whose authors had probably very different intents and background knowledge. The question is whether one can argue legitimately in such a fashion as this—a point given a certain piquancy by the fact that Rabbi Elliot Dorff, working from some of the same texts as the pope, argues that the Jewish tradition not only allows some abortions but under certain circumstances positively insists on them. The point is not that referring to the Bible is an illicit practice for believers—anything but—rather that it is often not easy to decide precisely what such reference implies and entails. It is for this reason that one should be wary of people who argue that philosophical discussion is unneeded for Christians and other believers. Even for them, there is need of reason and judgment. Faith unaided cannot do everything. And this is a point that apparently applies also to the third great religion that worships the God of Abraham—Islam. Abdulaziz Sachedina shows that Muslims differ over the state of the fetus and that they, too, have no easy answers. However, overall, he agrees with the more liberal Protestant Christians and the Jews that some stem cell research is not only morally and religiously permitted, but inasmuch as it is directed toward healing the sick and comforting the unfortunate, it is nigh mandated.

14
Testimony of
Nathan Salley

Nathan Salley for
National Bioethics Advisory Commission

Given July 17, 2001, United States House of Representatives Committee on Government Reform Subcommittee on Criminal Justice, Drug Policy, and Human Resources Hearing on Embryonic Cell Research

My name is Nathan Salley. I am sixteen years old and I will be a junior next year at Faith Christian Academy. I live with my parents, Mark and Leslie Salley, in Arvada, Colorado. My father and mother are with me today.

I am honored to represent some of the children that proponents of embryonic stem cell research insist they are trying to save. Yet embryonic stem cell research did not save me—cord blood research did. I am living proof that there are promising and useful alternatives to embryonic stem cell research and that embryos do not need to be killed to achieve medical breakthroughs.

My story begins at the age of eleven when I was ill for several months. My mother took me to doctors, who told me that I was the victim of tonsillitis, fatigue, and infections. They were dead wrong.

When—at my mother's urging—I was finally checked for mononucleosis, they found something much worse. The doctor who called with the results of my blood test told us to get to the hospital immediately.

On March 4, 1997, I was diagnosed with acute myloid leukemia. The disease was at an advanced stage by the time of diagnosis.

At age eleven, I knew nothing about leukemia. It was a terrible time, but I always tried to look on the bright side. Soccer teammates put my number "2" on their jerseys for the remainder of the season. Friends were very supportive, but the cancer treatment was awful. Cranial radiation exhausted me and chemotherapy caused me terrible nausea.

For eighteen months, I had chemotherapy for eighty-six to ninety-four hours each month and endured repeated spinal taps and bone marrow aspirations to check my progress. I lost my hair and appetite, but I tried hard to do as many things as I could—for life to be as normal as possible. Between chemotherapy treatments, I played soccer and kept up with school.

Since being diagnosed more than four years ago, I have spent nearly six months as an in-patient at the hospital and made nearly weekly visits to this day as an out-patient. I had to be homeschooled at the end of sixth grade, for all of seventh grade, for part of eighth grade and for part of my freshman year. I missed out on so many things with my friends that I could not begin to recount them. And, just when I thought the treatments were over and I was cured, I had a relapse.

Doctors informed me at age fourteen that I needed a bone marrow transplant. They gave me three options: receive the bone marrow from a donor relative, an unrelated donor, or cord blood. We found that nobody in my family was a match. We were ready to go ahead with a transplant from an adult donor who had what they call a five of six match with my proteins. However, at the last minute, a six of six matching cord blood unit from Spain became available.

Cord blood transplants were becoming more widely used by 1999, but were still experimental. Physicians assured us a cord blood transplant was my best chance for life and would reduce the likelihood of rejecting the transplant. After prayerful consideration, my parents and I followed their advice. Dad signed consent forms for me to participate in the procedure, requiring us to acknowledge that "umbilical cord blood transplantation has been performed mainly in small children and one of the purposes of this study is to determine whether it can be performed safely in larger people."

At fourteen, I was among the oldest children to receive a transplant from a small umbilical cord. More cells were going to be needed than were available in the cord blood unit. So the doctors told us about a second experimental procedure they felt should be used to expand the number of donated cord cells. We agreed to this procedure and signed more medical consent forms.

The scariest one said, "You are being asked to participate in a research study involving an investigational technique where a portion of your cord blood transplant will be treated in a laboratory before it is given back to you. The goal of this study is to determine whether we can speed up your marrow recovery by increasing the number of cord blood-forming cells in the laboratory before you receive them. . . . Your doctors have shown that if cord blood cells are treated in the laboratory for ten days with vitamins and growth factors, the number of stem cells which produce marrow recovery is substantially increased."

Before the transplant could take place my doctors had to completely kill my own leukemia-producing marrow with three days of total body radiation, followed by more intense chemotherapy. Then, the transplant took place in two phases. I received about 60 percent of the donated cord blood cells on June 29, 1999, when they arrived from Spain. The remaining cells were sent to the lab to be expanded. I was transfused with these cells ten days later on July 9, 1999.

It was an agonizing wait for my blood counts to begin to recover. I could not eat for two to three weeks. Thankfully, I am in complete remission today. Regular testing continues to show no leukemia present in my body. The transplanted cells have built a brand new bone marrow system for me.

When my transplant was performed by the doctors at Children's Hospital in Denver, I was just one of seven patients to receive a cord blood transplant in 1999. There have been just thirty-six total cord blood transplants at that hospital since the first one in 1996.

As a result of this groundbreaking procedure, I am proof that the medical community does not need to destroy life to save it. I am told that the same cord blood stem cells that saved me are likely cures for other life-threatening diseases. Differences of opinion exist about whether research using embryo stem cells may also yield medical benefits. But no one disputes that such research destroys embryos.

I am not a doctor, scientist, or theologian. And I am no more deserving than any other of the thousands of cancer patients and cancer survivors to have the opportunity to speak here. But, speaking as one cancer survivor who benefited from cord cell treatment, it does not seem right to me to terminate living human embryos based on mere speculation that they could lead to cures—when obvious alternatives are not yet exhausted.

All human life is "fearfully and wonderfully made." My life is no more valuable before God than the life of an embryo. Everyone wants to live a complete and healthy life, but I do not believe killing a life to save a life is right. Who, besides God, knows what people embryos may become? What we do know is that performing research on a four-day-old embryo will ensure that it never becomes a five-day-old embryo—much less a thirty-year-old actor, forty-year-old scientist who discovers a cure for Alzheimer's disease, fifty-year-old Congressman, or ninety-one-year-old former president.

Somehow the opportunity came to me from among countless others to be here today and tell my story. I have benefited from and participated in research on umbilical cord stem cells.

- Am I thankful to be alive today? Yes.
- Am I thankful that brilliant doctors and researchers discovered a cure for my disease? Absolutely.
- Would I want embryos unnecessarily killed when alternative research methods exist today? No.

So, I urge this committee—and President Bush—not to allow taxpayers' money to fund destruction of live human embryos.

Thank you.

15

Patients' Voices

The Powerful Sound in the Stem Cell Debate

Daniel Perry

Millions of patients may benefit from the applications of stem cell research, although there is disagreement about whether public funds should be used to develop the science. Patients have been key to winning political support. Acting as advocates, they have contended that public investment will speed the research and bring accountability to biomedical technology. A political dispute about the new research, which holds the potential for cures to devastating diseases and to foster healthy aging, shows the need to respect public sensibilities and to court public approval, as well as the importance of involving patients in debates where the methods of biomedical discoveries and ethical beliefs collide.

The achievement of isolating and growing cultures of self-renewing human pluripotent stem cells has set off waves of optimism among both researchers and the lay public.[1] The promise is tangible for effective new approaches to incurable diseases and underlying biological processes.[2] As shown in Table 1, over 100 million Americans suffer from illnesses that might be alleviated by cell transplantation technologies that use pluripotent stem cells. Yet some representatives in Congress and some of the lay public, as well as religious groups such as the National Conference of Catholic Bishops, oppose putting public funds behind the technology. They say that stem cell research belongs under a federal ban that currently prohibits federal funding of embryo research.[3]

PATIENTS FOR RESEARCH

In 1999, a coalition of three dozen national nonprofit patient organizations, the Patients' Coalition for Urgent Research (CURe), emerged to argue for public funding of human embryonic stem cell research under guidelines of the National Institutes of Health (NIH). This would achieve two goals: (1) participation by the broadest number of scientists under established peer-review mechanisms, thus rewarding the most promising research and speeding progress, and (2) public accountability and guidelines developed through processes that allow for public comment on an area of science that has raised ethical concerns.[4]

Why a patients' coalition? As taxpayers, patients and their family members are entitled to expect their government to make the most of a substantial public investment in biomedical research through the NIH and other agencies. And as the bearers of the ultimate burden when medicine cannot relieve their suffering, patients are the most compelling witnesses to the value of research that quite literally can save their lives.

In general, the patients and their advocates who are active for CURe display tempered optimism when it comes to appraising the chances of anyone's health benefiting soon from applications of stem cell research. Furthermore, broad views on the ethics and appropriateness of the technology have been expressed by those in CURe. For example, they believe in the principles of informed consent and free choice. Stem cell research must not lead to an underground black market in "spare" embryos for research. In addition, women and men, as individuals or as couples, should not be paid to produce embryos for research purposes.

The stories of patients and family members have fostered bipartisanship on Capitol Hill and have effectively complemented other activities such as the stance voiced by leading theologians from four major faiths—Roman Catholicism, Protestantism, Judaism, and Islam—who, noting the calls of their religions for compassion for the sick, wrote a joint letter to Congress urging federal involvement.[5]

THE BROADER STAKES

The promise of human pluripotent stem cell research increases the likelihood that vastly more people will experience healthy and productive aging. Age-related disease costs billions of dollars and burdens millions physically and financially.[6] The additional costs in medical and long-term care that are incurred annually in the United States because its Medicare recipients lose their functional independence are calculated at $26 billion.[7]

One can imagine the cost twenty years from now in the United States alone, when the population over age sixty-five is expected to double and the number of Americans over age eighty-five is projected to quadruple.[8] Unless bioscience engenders and receives broad popular support, in the future, nations like the United States, which have a rapidly increasing aging population, will more than likely struggle with a much greater health care burden. This is why it is so important to respect public sensibilities and to court public approval fervently, even though it is also likely that the next discoveries will, too, collide with the ethical and religious beliefs of some.

In the stem cell debate, patients have stepped forward to help draw the line between science in service to the community and science for lesser motives. Sadly, some of their most compelling stories will be silenced before long by the progression of their diseases. It surely behooves us to remember their contributions and to engage their successors, who will continue to put a human face on the promise of biomedical research.

Table 1. Persons in the United States Affected by Diseases That May Be Helped by Human Pluripotent Stem Cell Research. Data Are from the Patients' Coalition for Urgent Research, Washington, D.C.

Condition	Number of persons affected
Cardiovascular diseases	58 million
Autoimmune diseases	30 million
Diabetes	16 million
Osteoporosis	10 million
Cancer	8.2 million
Alzheimer's disease	4 million
Parkinson's disease	1.5 million
Burns (severe)	0.3 million
Spinal cord injuries	0.25 million
Birth defects	150,000 (per year)
Total	128.4 million

NOTES

1. J. A. Thomson et al., "Embryonic Stem Cell Lines Derived from Human Blastocycsts," *Science* 282 (1998): 1145; M. J. Shamblott, et al., "Derivation of Pluripotent Stem Cells from Cultured Human Primordial Germ Cell," *Proceedings of the National Academy of Sciences of the United States of America* 95 (1998): 13726.

2. American Association for the Advancement of Science and Institute for Civil Society, *Stem Cell Research and Applications: Monitoring the Frontiers of Biomedical Research*, November 1999.

3. Rep. J. Dickey (R-Ark), "No Such Thing as Spare Embryos," *Roll Call* (3 June 1999), p. 4. R. M. Doerflinger, testimony on behalf of the Committee for Pro-Life Activities of the National Conference of Catholic Bishops before the Senate Appropriations Subcommittee on Labor, Health, and Education hearing on legal status of embryonic stem cell research (Senate Hearing 105-939), 26 January 1999 [online], frwebgate.access.gpo.gov/cgi-bin/getdoc.cgi?dbname5105_senate_hearings&docid5f:54769.wais.

4. Goals adopted by Patients' CURe, Washington, D.C., 20 May 1999.

5. "Theologians from Four Major Faiths Express Support for Federal Funding of Stem Cell Research," press release from Patients' CURe, Washington, D.C. (14 October 1999).

6. Alliance for Aging Research, *A Call for Action: How the 106th Congress Can Achieve Health and Independence for Older Americans Through Research*, 1999.

7. Alliance for Aging Research, *Independence for Older Americans: An Investment for Our Nation's Future. A Report by the Alliance for Aging Research*, 1999.

8. Ibid.

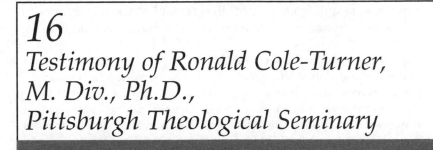

16
Testimony of Ronald Cole-Turner, M. Div., Ph.D., Pittsburgh Theological Seminary

National Bioethics Advisory Commission

I want to thank the commission for the attention that you are giving today to religious perspectives regarding human stem cell research and for the opportunity to speak before you. I come as a member of a mainline Protestant denomination, the United Church of Christ, and although no one individual speaks for our church, I will try to represent the positions we have taken and the concerns that we hold.

Let me begin by saying that we do not have an official position regarding the status of embryos. That is not to say we have no opinion or do not care about their rightful status before God. But officially, we have never declared that we regard embryos as persons. Some of our members would agree with that declaration; many—perhaps most—would not agree, believing instead that embryos have an important but lesser status. But we have, deliberately, I think, avoided any such declarations. On the contrary, we have made statements in which we express our openness to embryo research, given certain conditions, which I will come to in a moment.

I quote at length from a report that served as the background to a 1997 General Synod resolution on the question of human cloning:

> Beginning with the 8th General Synod in 1971, various General Synods of the United Church of Christ have regarded the human pre-embryo as due great respect, consistent with its potential to develop into full human personhood. General Synods have not, however, regarded the pre-embryo as the equivalent of a person. Therefore, we on the United Church of Christ Committee on Genetics do not object categorically to human pre-embryo research, including research that produces and studies cloned human pre-embryos through the fourteenth day of fetal development, provided the research is well justified in terms of its objectives, that the research protocols show proper respect for the pre-embryos, and that they are not implanted. We urge public discussion of current research and future possibilities, ranging from pre-implantation genetic screening of human pre-embryos to nuclear transfer cloning to human germline experimentation. We do not categorically oppose any of these areas of research, but we believe they must be pursued, if at all, within the framework of broad public discussion. In 1989, the 17th General Synod of the United Church of Christ stated that it was "cautious at present about procedures that would make genetic changes which humans would transmit to their offspring (germline therapy). . . . We urge extensive public discussion and, as appropriate, the development of federal guidelines during the period when germline therapy becomes feasible." . . . . We on the United Church of Christ Committee on Genetics are opposed to the idea that human pre-embryo research, such as germline experimentation or research involving cloned pre-embryos, should be permitted but left largely unregulated if funded privately, or that there is no federal responsibility for the ethics of such research if federal funds are not used. We believe that this approach merely seeks to avoid the difficult public deliberation that should occur prior to such research. We believe that all such research should be subject to broad public comment and that it should only proceed within a context of public understanding and general public support.[1]

And so when it comes to the specific questions before you regarding the ethics of pluripotent human stem cell research and federal policy in this area, my view is that it is broadly consistent with the views of the United Church of Christ that human stem cell research go forward with federal funds. In fact, we go further and encourage reconsideration of the ban on federal funding for embryo research. We are open to the possibility that somatic cell nuclear transfer be used to create embryos for research, but not implantation, under highly defined research protocols, and that this research, too, be done with public funding.

One of the conditions that we attach to the possibility of this research is that a clear and attainable benefit for science and for medicine be indicated in advance. And it is reasonable to think that now, with pluripotent stem cell technology, such benefit is becoming clearer.

Another condition that we attach is that this research should follow a period of intense and open public discussion. In fact—and let me be as clear as I can about this—all that I have said about our support for research in these areas depends upon meeting the condition of advanced public discussion. I believe this is especially important for this commission, because it represents one of very few places in our national life where such a conversation can begin.

We stipulate this condition for two reasons. First, we believe that although enormous advances for medicine lie ahead in these areas of research and that we are obliged to work to achieve these advances, our efforts could be undermined, and it could be very bad for science if research proceeds in the short term without broad public understanding and support. Public misunderstanding and public exclusion from discussion could result in public rejection of this and related forms of research.

The second reason why we set forth the condition of advanced public discussion and support is that we value living in a society that makes basic public moral decisions based on the deliberations of informed citizens. As a historic church, our congregational forebears extended congregational decisionmaking to the public square. As a church today, we believe that our views are not the only views worth hearing, but that public policy on morally problematic issues should be the result of honest and sustained discourse during which all views are brought forward in public. This view of a public society is an article of faith with us.

As a commission, you are, of course, under time constraints and must offer your report on specific policy questions. As a church, we offer at least some support for the view that federally funded research in embryonic stem cells, and possible even in embryos, should move forward as quickly as possible. But on the basis of the condition our Church has set on this support, I ask you to do whatever you can in your report to satisfy our condition by helping to bring about a new, open, and sustained national discussion of these difficult questions and issues. Such a sustained discussion may be well beyond your mandate and may require some new institutional platform, but you are one of the key voices in our national life that can urge that this challenge be taken on for the good of research, for the good of public support of research, and for the good of the kind of society in which we want to live.

I will conclude by noting two concerns, both of which involve contextual factors that a church such as ours will bring to the discussion table that I am urging you to help create. The first is social justice. Precisely because this research promises such great benefit, we worry that the benefit will be distributed unevenly and therefore that it will further the position of the rich and the powerful at the expense of the poor and the weak. We believe that the moral test of any system, including our system of medical research and treatment, is how well it treats the least privileged members of society, first of all within our own nation, but also globally. And so we would challenge those who fund and develop these therapies by asking the following question: How will the benefits be shared universally? We are aware that the difficult problems of delivery and cost recovery must be considered, but in offering our support for this research based on the promise of medical benefit, we do not mean that the benefit should be distributed only by market means.

The second concern involves the broader scientific and medical context of research. It is impossible for any of us to offer a moral assessment of human stem cell technology in isolation from other current or pending areas of research, among them somatic cell nuclear transfer and human germline modification. Through these technologies and the combination of these technologies, we are about to acquire a wholly unprecedented level of control over our health, our longevity, and our offspring. And so I urge you to do whatever is in your power not only to create a broad public discussion, but also to define its agenda broadly as involving this wide but inter-related set of emerging technologies.

I conclude with this simple observation: If the question before us is narrowly defined as involving embryos and stem cells, the various religious traditions will take different positions. But if the question is framed in terms of concern for social justice or of our ability to chart our common future in view of the overwhelming changes that lie ahead, the various religious traditions will find there is much upon which to agree. If this is correct, then we might find greater understanding on the narrow issues as we move along the pathway of greater engagement on the contextual issues.

NOTE

1. "'Statement on Cloning' by the United Church of Christ Committee on Genetics," in *Human Cloning: Religious Responses*, ed. Ronald Cole-Turner (Louisville, Ky: Westminster John Knox Press, 1997), pp. 149–51.

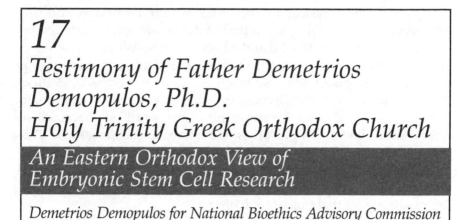

17
Testimony of Father Demetrios Demopulos, Ph.D.
Holy Trinity Greek Orthodox Church

An Eastern Orthodox View of Embryonic Stem Cell Research

Demetrios Demopulos for National Bioethics Advisory Commission

I would like to thank the commission for providing me with an opportunity to present an Orthodox view of the ethical problems and challenges associated with human embryonic stem cell research. I would like to emphasize that I do not speak for the Greek Orthodox Church but instead offer comments that I believe are consistent with the teachings and tradition of the Orthodox Church.

The Orthodox Church has a long tradition of encouraging the "medical art" that alleviates unnecessary pain and suffering and restores health. The Church, however, also has reminded us that this art is given to us by God to be used according to His will, not our own, since "the medical art has been vouchsafed us by God, who directs our whole life, as a model for the cure of the soul" and "we ought not commit outrage against a gift of God by putting it to bad use."[1] What constitutes bad use is what has brought us together here today. An important consideration for the Orthodox is based on our understanding of what it is to be a human person.

Humans are created in the image and likeness of God and are unique in creation because they are psychosomatic, beings of both body and soul—physical and spiritual. We do not understand this mystery, which is analogous to that of the theanthropic Christ, who at the same time is both God and a human being. We do know, however, that God intends for us to love Him and grow in relationship to Him and to others until we reach our goal of theosis, or deification, participation in the Divine Life through His grace. We grow in the image of God until we reach the likeness of God. Because we understand the human person as one who is in the image and likeness of God, and because of sin we must strive to attain that likeness, we can say that an authentic human person is one who is deified. Those of us who are still struggling toward theosis are human beings, but potential human persons.[2]

We believe that this process toward authentic human personhood begins with the zygote. Whether created *in situ* or *in vitro*, a zygote is committed to a developmental course that will, with God's grace, ultimately lead to a human person. The embryo and the adult are both potential human persons, although in different stages of development. As a result, Orthodox Christians affirm the sanctity of human life at all stages of development. Unborn human life is entitled to the same protection and the same opportunity to grow in the image and likeness of God as are those already born.

Given this Orthodox understanding of human personhood and life, I cannot condone any procedure that threatens the viability, dignity, and sanctity of that life. In my view, the establishment of embryonic stem cell lines[3] was done at the cost of human lives. Even though not yet a human person, an embryo should not be used for or sacrificed in experimentation, no matter how noble the goal may seem. For me, then, the derivation of embryonic stem cell lines is immoral because it sacrificed human embryos, which were committed to becoming human persons. That the embryos donated for this work were not going to be implanted and had no chance of completing their development cannot mitigate the fact that they should not have been created. *In vitro* fertilization techniques that routinely result in "surplus" embryos that are eventually discarded is immoral for the same reasons I have already mentioned. I believe, then, that the prohibition of research using human embryos should be continued and, if possible, extended to the private sector as well.

Wishing that something had not been done will not undo it. Established embryonic stem cell lines exist, and their use has great potential benefits for humanity, which need not be reviewed here. The Orthodox Church, as I mentioned before, has a long tradition of encouraging the medical arts. We have a long list of healer-saints— physicians who became authentic persons through the practice of medicine. Invariably, they obeyed the commandment of Christ to his apostles, "Heal the sick, raise the dead, cleanse lepers, cast out demons. You received without paying, give without pay."[4] Without going into an extensive exegesis of the verse, the intention is clear: Attend not to profit, but to the medical needs of others.

Using our healer-saints as a paradigm, I am concerned about how the existing stem cell lines will be used. Will they be used to heal, or will they be used to maximize profits? Market forces are very strong, and, in my opinion, are often contrary to the general good. Allowing the cell lines to be owned by private companies that are responsible first to their stockholders and investors rather than to the general welfare may compromise the use of the lines. It is imperative that steps be taken to ensure that the lines be used only for therapeutic procedures that will benefit those in need and not be limited to the few who will be able to afford them. I want to emphasize that the lines must be used only therapeutically, to restore health and to prevent premature death. They must not be used cosmetically or to further any eugenic agenda. None of us is physically perfect, but all are called to be perfected in Christ. Part of our challenge to participate in the Divine Life is to overcome our deficiencies. We must not attempt to re-create ourselves in our own image.

Because stem cell lines have such great potential for healing, efforts should be made to encourage discovery of more morally acceptable sources. A recent report suggests that adult stem cells may be less restricted than previously thought.[5] It may be possible to develop techniques to culture such cells without the need to sacrifice the donor. Alternatively, because organ donation is viewed favorably by many (but not all) Orthodox Christians, I would accept cell lines derived from fetal primordial germ cells, but only in cases of spontaneous miscarriage. A fetus cannot be killed for an organ, just as an adult cannot. Also, great care must be taken to assure that the mother's consent is truly informed.

In summary, the Orthodox Church promotes and encourages therapeutic advances in medicine and the research necessary to realize them, but not at the expense of human life. The Church considers human life to begin with the zygote and to extend beyond our physical death, as we were promised eternal life by our God and Savior. Recognizing that we are all in a sinful and imperfect state, the Church admonishes us to strive for perfection through God's grace as we strive to become authentic human persons in communion with God. Because we tend to follow our own will rather than God's, we are reminded to be discerning so that we do not commit outrages by putting a gift of God to bad use.

NOTES

1. St. Basil the Great, *The Long Rules* 55, in *St. Basil: Ascetical Works, The Fathers of the Church*, vol. 9, trans. M. M. Wagner (Washington, D.C.: Catholic University of America Press, 1962), pp. 330–37.

2. See also, P. Nellas, *Deification in Christ: The Nature of the Human Person* (Crestwood, N.Y.: St. Vladimir's Seminary Press, 1987), ch. 1 and 2 and J. Breck, *The Sacred Gift of Life: Orthodox Christianity and Bioethics* (Crestwood, N.Y.: St. Vladimir's Seminary Press, 1998), ch. 1 and 3 .

3. J. A. Thomson, et al., "Embryonic Stem Cell Lines Derived from Human Blastocysts," *Science* 282 (1998): 1145–47.

4. Matt. 10:8.

5. C. R. Bjornson, et al., "Turning Brain into Blood: A Hematopoietic Fate Adopted by Adult Neural Stem Cells *In Vivo*," *Science* 283 (1999): 534–37.

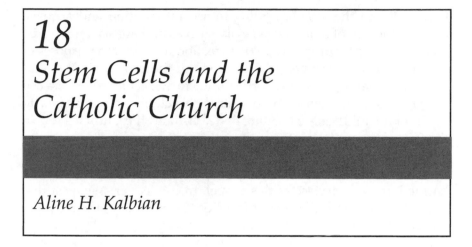

18
Stem Cells and the Catholic Church

Aline H. Kalbian

Strong opposition and condemnation by the Catholic Church is a prominent feature of most contemporary debates about cloning and the use of human embryos in research. So prominent, in fact, that while President Bush was pondering the fate of federal funding for embryonic stem cell research in July of 2001, he met with Pope John Paul II at the Vatican and was urged by him to "reject practices that devalue and violate human life at any stage from conception until natural death."[1] The president's consultation with a religious leader on this issue was certainly remarkable, especially to those concerned with church/state separation. Equally remarkable, however, has been the Church's resolve on this issue.

The particulars of the Church's position on embryonic stem cell research often meld together with a larger set of views opposing abortion, artificial contraception, and most forms of assisted reproductive technology. In fact, in the Pope's comments to President Bush, he explicitly groups several activities under the label of "related evils," asserting that "Experience is already showing how a tragic coarsening of consciences accompanies the assault on innocent

human life in the womb, leading to accommodation and acquiescence in the face of other related evils, such as euthanasia, infanticide,
and most recently proposals for the creation for research purposes of
human embryos, destined to destruction in the process."[2] This slippery slope that begins with abortion leads to evil actions like stem cell
research using human embryos. All such actions are connected to
what John Paul II calls a "culture of death," and the Church's role in
the face of this culture is to proclaim the "gospel of life."

The embryo's status as a human person entitled to full human
rights from the moment of conception, including the right not to be
harmed, is a centerpiece of this gospel. In this paper, I suggest that
while the embryo *is* the central focus of the argument, the image of
the embryo represents something much larger; namely a certain
understanding of the vulnerable *other*, the neighbor in society. References to the embryo serve as a metaphor that reveals the central
Christian message of neighbor love. However, this metaphorical language also hides a set of beliefs about procreation and family. I begin
with a brief exploration of how the embryo as metaphor functions. I
then set the teaching about the status of the embryo in the broader
context of the Catholic Church's view on embryonic stem cell
research. Finally, I look at the extent to which the Catholic position is
unchanging and absolute. I do this by outlining three elements of
Catholic moral reasoning: natural law, the common good, and proportionalism. These three elements all reveal tensions in contemporary Catholic moral thought—tensions that can provide an opening
for a different analysis of the use of human embryos in research.

THE EMBRYO AS NEIGHBOR

Pope John Paul II's writings on morality have long echoed the
theme of caring for the vulnerable other. In recent decades the image
of the embryo has served as the most effective means for capturing
the vulnerability of the other who needs protection. On this view,
society's acceptance of abortion is intimately related to a larger rejection of "a truly humane future." What becomes clear in much of
John Paul II's writing is his view that the Catholic Church ought to
speak for those who have no voice or who lack the capacity to speak
for themselves.

In *Evangelium Vitae* (EV), the pope's 1995 encyclical on abortion and euthanasia, he describes this care and concern for the other in terms of the "positive requirements" that emerge from the "commandment regarding the inviolability of life."[3] According to EV, the prohibition against killing is a negative commandment; its positive face is the commandment to love the neighbor. As John Paul II reminds his readers, the category neighbor includes not only friends and close relations, but also strangers and enemies. All are neighbors because they are made in the image of God. Thus, the attitude of love *for* the other and the prohibition *against* killing are all part of the same commandment. Interestingly, the attitude of love and reverence for life is intensified from mere attitude to action when the Pope encourages the faithful "to defend and promote life." The ultimate expression of this defense and promotion of life is God's call to humans to "be fruitful and multiply."

The activity of procreation is intimately tied to the protection and defense of life. Just as the neighbor love command is the positive manifestation of the prohibition against killing, procreation is the positive action that correlates with protecting and defending life. Human procreation involves the "special participation of man and woman in the creative work of God."[4] Humans, on this view, cooperate with God who is understood to be present at the moment of conception. This is based on the idea that the new life bears the image and likeness of God. John Paul II writes, "the genealogy of the person is inscribed in the very biology of generation."[5] That genealogy, marked through the *imago dei*, confirms the status of the embryo as person made in God's image. In essence, then, harm to the embryo at any stage is harm to this image and destruction of the embryo destroys the genealogy. Humans as cocreators with God have a special responsibility or mission that requires "accepting and serving life" especially life "when it is at its weakest."[6] This weakness and vulnerability is especially evident in human life at its beginning and its end. Thus, the embryo is in special need of protection.

In the Vatican Instruction, *Donum Vitae* (DV), the Catholic Church asserts the view that while parents do not have a "right" to a child, the child (from the moment of conception) does have a right to "be conceived, carried in the womb, brought into the world, and brought up within marriage."[7] This attribution of rights to the zygote, embryo, and fetus emphasizes the strong connection between marriage and procreation in the tradition. The clear subtext is that the married couple takes on the task of procreation as part of the battle to defend and promote life.

THE CATHOLIC ARGUMENT AGAINST EMBRYONIC STEM CELL RESEARCH

Concern about the treatment of human embryos has long been at the center of Catholic views about reproductive technologies and abortion. The Church has held the unequivocal stance that "Life once conceived, must be protected with the utmost care."[8] Thus, the recent explicit responses to cloning and stem cell research are not new.

The Catholic literature about embryonic stem cell research shares certain salient characteristics. Five are worth noting. The first, and in many ways most significant, is that the embryo from the first moment of fertilization is a human subject with an identity. The most recent Vatican statement on stem cell research identifies this as the primary reason why it is morally illicit to produce and/or use human embryos for the purpose of deriving stem cells. The basis for this attribution of subjectivity and "well-defined identity" is said to be "a complete biological analysis" which leads to the conclusion that from the moment of conception the human embryo "begins its own coordination, continuous and gradual development."[9] This view has been articulated in a broad range of Catholic documents. For example, in the 1987 Vatican Instruction, *Donum Vitae*, we read the following statement: "This teaching remains valid and is further confirmed, if confirmation were needed, by recent findings of human biological science which recognize the zygote (the cell produced when the nuclei of the two gametes have fused) resulting from fertilization the biological identity of a new human individual is already constituted."[10] This quasi-scientific argument is supplemented in Catholic teaching by the belief in the totality or unity of body and spirit. Thus from the very moment of conception, an individuated material entity exists. Since this entity (the newly formed zygote) exists materially as a body, the argument goes, it must also exist as a spirit, and is thus a full human individual.

The notion of an identity is expressed in countless ways through many of the Vatican documents. In the encyclical *Evangelium Vitae*, Pope John Paul relies on biblical passages for evidence of the subjectivity of the embryo. One of the most often quoted passages, "Before I formed you in the womb I knew you, and before you were born I consecrated you"[11] is mentioned but not explicated in much detail. It is presented as proof that all human beings, even while still in their mother's womb, are "the personal objects of God's loving and fatherly providence"[12] and that "the life of every individual from its very beginning, is part of God's plan."[13] More interesting and elaborate are the references to Job who in the midst of his despair is able to know that "there is a divine plan for his life."[14] Job's trust in God is based partly on his sense of awe at God's creative work. Job's meditation on himself as God's creation is filled with detail. He asks, "Did you not pour me out like milk and curdle me like cheese? You clothed me with skin and flesh, and knit me together with bones and sinews."[15] Of course, this lament is part of an appeal on Job's part for God's mercy. By reminding God that He is the creator who fashioned him, Job is pleading with God. The use of this passage by the Pope as a proof-text that the fetus in the womb is a subject with an identity is interesting because the passage makes no mention of the womb or the moment of conception. It simply affirms God's creative power.

The other passage used by the Pope is the story from the gospel of Luke of the meeting between Elizabeth and Mary who are both pregnant. The significant moment in the story occurs when Mary's greeting to Elizabeth causes Elizabeth's fetus to "leap in her womb." On the Pope's reading, this activity in the womb affirms that "it is precisely the children who reveal the advent of the Messianic age."[16] This evidence of life in the womb is then translated into confirmation of "the indisputable recognition of the value of life from its very beginning."[17] Quoting St. Ambrose of Milan, the Pope highlights the role of "the babies" (John the Baptist and Jesus Christ): "The women speak of grace: The babies make it effective from within to the advantage of their mothers who by a double miracle prophesy under the inspiration of their children."[18] The children or fetuses are exalted and imbued with an almost supernatural power; the mothers are made to seem mere tools in this important transmission. This image is especially effective in the Catholic argument against abortion. It responds to any arguments that characterize abortion as the mother's life vs. the life of the fetus by affirming the view that the fetus takes precedence.

Once the identity of the embryo/fetus is confirmed the next element of the Catholic view follows. It is the philosophical point that this newly formed human subject has rights; namely a right to its own life. In *Evangelium Vitae*, the Pope writes "it must nonetheless be stated that the use of human embryos or fetuses as an object of experimentation constitutes a crime against their dignity as human beings who have a right to the same respect owed to a child once born, just as to every person."[19] While clearly not an argument, this statement emphasizes the sense that at all stages of development, the embryo is a bearer of rights. Thus, from identity as a subject, the move is made to assign full personhood to the embryo. In the language of the Catholic Church the embryo is categorically a human being and a rights-bearing person.

Once the status of the embryo has been determined, both on the basis of biological and philosophical reasoning, the Church then claims that harming the embryo by "irremediably damaging" it or by "curtailing its development is a gravely immoral act and consequently is gravely illicit."[20] This is the moral claim, and it is intricately wound up in the Church's Vatican II teaching about the interdependence of the individual person and the larger community. In *Gaudium et Spes* we see the articulation of this view in the statement, "Man's social nature makes it evident that the progress of the human person and the advance of society itself hinge on each other. For the beginning, the subject and the goal of all social institutions is and must be the human person. . . ."[21] Thus, regardless of the larger benefit that might accrue from stem cell research on embryos, the individual embryo is one and the same as "the human person" for whom social institutions exist.

The fourth element, which is essentially the core of the justification for the above-mentioned moral claim, is stated in the following dictum: A good end cannot justify and evil means. This anticonsequentialist strain in the Catholic tradition serves to rebut the sorts of arguments that weigh the benefits of curing Alzheimer's, Parkinson's, diabetes, and other diseases against the sacrifice of human embryos that have not yet been implanted in a woman's womb. Put differently, the action of harming or destroying an embryo is inherently wrong and its wrongness cannot be lessened in any way by the potential therapeutic procedures that may result.

The fifth and final element of the Catholic argument provides a context by stressing its historical consistency and longevity. This affirmation of history and tradition is an important feature of Catholic moral theology, and is evident in most magisterial documents. The consistency of this teaching serves to remind all that it is not open to revision. This question of the unchanging nature of Catholic doctrine, especially on matters of morality, has recently received quite a bit of attention from Catholic moral theologians. I turn to that in the next section.

CAN CATHOLIC TEACHING ON EMBRYONIC STEM CELL RESEARCH CHANGE?

While this question cannot be responded to easily, it is possible to point to areas of tension in recent debates about Catholic moral theology. Three areas in particular reveal fissures where it might be possible for critics of the tradition to find ways to allow for certain types of embryonic stem cell research. These three areas are (1) critiques of natural law, (2) the rise of social justice arguments, and (3) the legacy of proportionalism.

Natural Law

Natural law theory is practically synonymous with Catholic moral theology, especially with respect to teachings on moral issues related to the body. The belief that human reason's reflections on nature and experience lead to moral knowledge is based on scripture[22] and on the view of the goodness of God's creation. This strong confidence in human reason is evident in the conclusion that natural law enables all persons, not only Catholics, access to moral wisdom and truth. It also affirms the coparticipation of humans in the Divine Law. Human participation in God's plan for creation involves humans in the very act of creation. This emphasis echoes the image of the man and woman's special participation in God's creative work through the event of procreation.

In the decades since Vatican II, a revisionist strain has emerged in Catholicism, and much of its efforts have been directed at revising natural law. This revision has taken a variety of forms. The main problem that many contemporary Catholic theologians who work in medical ethics have noted about traditional natural law is what has been dubbed its physicalist nature. Physicalism refers to the belief that the human body's tendencies are natural and thus from observing these tendencies we can derive moral norms.

Physicalism is best described using the terms—"order of nature" and "order of reason." "Order of nature" refers to the strain of interpretation of the natural law that focuses "on the physical and biological structures given in nature as the source of morality."[23] The other strain of natural law interpretation, "the order of reason" emphasizes natural law as the "human capacity to discover in experience what befits human well-being."[24] Interestingly, while these are identified as two separate strains, many scholars find evidence that Thomas Aquinas accepted both, and as a consequence, claim that contemporary Catholic moral theology ought to incorporate both. This distinction between the two strains of natural law can also be described as the biological vs. the rational strain. The physicalist critique of natural law is directed at an overemphasis or overvaluation of the biological strain. The claim is that deriving moral norms from physical and biological structures and functions violates the totality (unity of body and spirit) of the person. It isolates the corporeality of the person—viewing her just as a body. The critics of an overly physicalist natural law cite the Church's absolute prohibitions on contraception and artificial insemination by husband as two examples of natural law that privileges the biological over the rational.

Pope John Paul II responds to this attack on natural law in contemporary Catholic teaching in his 1993 encyclical letter *Veritatis Splendor*. His response combines the two orders mentioned above into a unified whole which he describes as "man's proper and primordial nature," the "nature of the human person," which is the *"person himself in the unity of soul and body,* in the unity of his spiritual and biological inclinations and of all the other specific characteristics necessary for the pursuit of his end."[25] He wants to retain an emphasis on the moral relevance of the body, but by speaking of the body only in terms of unity and totality, he hopes to avert the accusation of physicalism.

The relevance of this discussion for stem cell research on human embryos is twofold. First, critics of contemporary Catholic teaching can claim that a "physicalist" application of the natural law distorts the full context of embryonic stem cell research. Second, defenders of the tradition can respond using the totality of the body and spirit as evidence that the embryo, even from its earliest stages, is due respect.

Social Justice

The second issue that illuminates this debate is the emphasis in the twentieth century on social justice arguments in Catholic moral theology, especially as they incorporate the notion of the common good. The common good is defined in the Second Vatican Council documents as "the sum of those conditions of social life which allow social groups and their individual members relatively thorough and ready access to their own fulfillment."[26] The sections of *Gaudium et Spes*, the Vatican II *Pastoral Constitution on the Church in the Modern World*, that discuss the common good exhibit the delicate relationship between the individual and society. The Constitution states that "the social order and its development must unceasingly work to the benefit of the human person."[27] The central point is that the social order must be subordinate to the dignity of the individual. This point is central to the Catholic opposition to embryonic stem cell research in the sense that even if the social order could be improved by the use of embryos in research, the assault on the individual's life and dignity (in this case the human embryo) cannot be justified. Nevertheless, it might be possible for some Catholics, especially those who question the full personhood of the embryo, to claim that the common good is benefited by embryonic stem cell research.

Proportionalism

The third interesting element to consider is the debate about proportionalism. Proportionalism is the term used to describe a movement of revisionist theologians who emerged in both the American and European contexts on the heels of the Second Vatican Council in the 1960s. Their primary contention is that the "principle of proportionate reason," which asserts that a proportionality must exist between the means and the end of an action, is an integral part of traditional Catholic moral theology. Essentially, though, the larger project of revision entails a re-evaluation of the object of the moral act. They argue for more attention to the context (both intentions and consequences) of moral actions. Put differently, they claim that the means for achieving an end must be proportionate to that end. Thus, the object of the act, which proportionalists claim has been construed too narrowly by the magisterium, necessarily must include this sense of proportion.

As one might imagine, proportionalism has not been embraced by the magisterium, which has claimed that proportionalism is a form of utilitarianism, despite the fact that proportionalists see themselves as squarely within the tradition of Catholic moral theology. Nevertheless, some have argued that its contributions are more significant than such a dismissal might suggest. Proportionalist approaches would not necessarily lead to full support of cloning and embryonic stem cell research. The call for greater attention to context might, however, provide an opening for Catholics who want to allow for some use of embryonic stem cells and still remain faithful to the tradition.

CONCLUSION

One final issue that perplexes many observers of the Church is its response to research using stem cells from embryos that have already been destroyed. Many find it difficult to place moral blame on the scientist who had nothing to do with the destruction of the embryo and is now using the material for a good end. According to the Church, while destruction and use of embryos are two distinct moral acts, both are morally illicit. In the preceding discussion I focused on the argument directed at the destruction of the embryos. In cases when the embryo has already been produced (for example extra embryos resulting from *in vitro* fertilization procedures), the moral issue becomes one of cooperation or complicity with evil. The Catholic tradition has a long history of commentary on the question of complicity and cooperation. According to the most recent edition of the Catechism, cooperation in the sin of others includes direct and voluntary participation; praise and approval; failure to disclose or hinder the evil doer; or protection of the evil doer. Thus, any involvement with biological materials from illicitly obtained embryos is also the responsibility of the cooperating individual.

Catholic documents and statements on cloning, stem cell research, and reproductive technologies all attempt to walk a fine line in their depiction of science and technology. They affirm scientific endeavors that will result in appropriate therapeutic applications, but they want to insure that science and technology are always seen to be in the service of humanity. In the view of the Church, when embryos become tools or objects of scientific study, not only has humanity harmed those particular embryos, but it has also assaulted all that is associated with the vulnerable and weak neighbor.

NOTES

I would like to thank Thomas J. Neal for help in gathering the sources for this paper and Robert R. Cross and John Kelsay for suggestions and comments on an earlier draft.

1. Pope John Paul II, "To the President of the United States of America, H. E. George Walker Bush," [online], www.vatican.va/holy_father/john_paul_ii/speech/2001/july/index.htm [23 July 2001].

2. Ibid.

3. Pope John Paul II, *Evangelium Vitae* (*The Gospel of Life*) (New York: Random House, 1995), p. 41.

4. Ibid., p. 43.

5. Ibid.

6. Ibid.

7. Congregation for the Doctrine of Faith, "Donum Vitae (Instruction for Human Life in its Origin and the Dignity of Procreation): Replies to Certain Questions of the Day," [online], www.vatican.va/roman_curia/congregations/cfaith/index.htm [February 22, 1987]. See II, A, 2.

8. Ibid., I, 1.

9. Pontifical Academy for Life, "Declaration on the Production and the Scientific and Therapeutic Use of Human Embryonic Stem Cells," [online],www.cin.org/docs/stem- cell-research.html [25 August 2000], p. 4.

10. Congregation for the Doctrine of Faith, p. 701.

11. Jere. 1:5.

12. Pope John Paul II, *Evangelium Vitae*, p. 61.

13. Ibid., p. 44.

14. Ibid.

15. Job 10:10–11.

16. Pope John Paul II, *Evangelium Vitae*, p. 45.

17. Ibid.

18. Ibid.

19. Ibid., p. 63.

20. Pontifical Academy for Life, p. 4.

21. Pope Paul VI, "*Gaudium et Spes* (Pastoral Constitution on the Church in the Modern World)," [online], www.vatican.va/archive/hist_councils/ii_vatican_council/index.htm [7 December 1965], p. 25.

22. Rom. 2:14–16.

23. Richard M. Gula, S.S. *Reason Informed by Faith: Foundations of a Catholic Morality* (New York: Paulist Press, 1989), p. 223.

24. Ibid.

25. Pope John Paul II, *Veritatis Splendor* (*The Splendor of Truth*) (Boston: St. Paul Books and Media, 1993), p. 50.

26. Pope Paul VI, "*Gaudium et Spes*," p. 26.

27. Ibid.

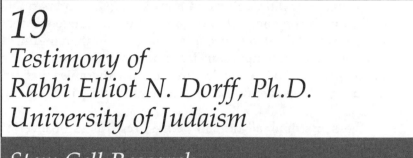

19
Testimony of Rabbi Elliot N. Dorff, Ph.D. University of Judaism

Stem Cell Research

Elliot N. Dorff for National Bioethics Advisory Commission

FUNDAMENTAL THEOLOGICAL CONVICTIONS[1]

1. The Jewish tradition uses both theology and law to discern what God wants of us. No legal theory that ignores the theological convictions of Judaism is adequate to the task, for any such theory would lead to blind legalism without a sense of the law's context or purpose. Conversely, no theology that ignores Jewish law can speak authoritatively for the Jewish tradition, for Judaism places great trust in law as a means to discriminate moral differences in similar cases, thus giving us moral guidance. My understanding of Judaism's perspective on stem cell research will, and must, draw on both theological and legal sources.

2. Our bodies belong to God; we have them on loan during our lease on life. God, as owner of our bodies, can and does impose conditions on our use of our bodies. Among those conditions is the requirement that we seek to preserve our lives and our health.

3. The Jewish tradition accepts both natural and artificial means

for overcoming illness. Physicians are the agents and partners of God in the ongoing act of healing. Thus, the mere fact that human beings created a specific therapy rather than finding it in nature does not impugn its legitimacy. On the contrary, we have a duty to God to develop and use any therapies that can aid us in taking care of our bodies, which ultimately belong to God.

4. At the same time, all human beings, regardless of level of ability and disability, are created in the image of God and are to be valued as such.

5. Moreover, we are not God. We are not omniscient, as God is, and so we must take whatever precautions we can to ensure that our actions do not harm ourselves or our world in our very effort to improve them. A certain epistemological humility, in other words, must pervade whatever we do, especially when we are pushing the scientific envelope, as we are in stem cell research. We are, as Genesis says, supposed to work the world and preserve it; it is the achievement of that balance that is our divine duty. [2]

JEWISH VIEWS OF GENETIC MATERIALS

1. Because doing research on human embryonic stem cells involves procuring them from aborted fetuses, the status of abortion within Judaism is a subject that immediately arises. Within Judaism, by and large, abortion is forbidden. The fetus, during most of its gestational development, is seen as "the thigh of its mother," and neither men nor women may amputate their thigh at will, because that would be injuring their bodies, which belong to God. On the other hand, if the thigh turns gangrenous, both men and women have the positive duty to have their thigh amputated in order to save their lives. Similarly, if a pregnancy endangers a woman's life or health, an abortion must be performed to save her life or protect her physical or mental health, for she is without question a full-fledged human being with all the protections of Jewish law, while the fetus is still only part of the woman's body. When there is an elevated risk to the woman beyond that of

normal pregnancy, but insufficient risk to constitute a clear threat to her life or health, abortion is permitted, but it is not required. That is an assessment that the woman should make in consultation with her physician. Some recent authorities also would permit abortions in cases where genetic testing indicates that the fetus will suffer from a terminal disease such as Tay-Sachs or from serious malformations.[3]

The Jewish stance on abortion, then, is that if a fetus was aborted for legitimate reasons under Jewish law, it may be used to advance our efforts to preserve the life and health of others. In general, when a person dies, we must show honor to God's body by burying it as soon as possible after death. To benefit the lives of others, however, autopsies may be performed when the cause of death is not fully understood, and organ transplants are allowed to enable other people to live.[4] The fetus, as I have said, does not have the status of a full-fledged human being. Therefore, if we can use the body of a human being to enable others to live, how much the more so may we use a part of a body—in this case, the fetus—for that purpose. This all presumes that the fetus was aborted for good and sufficient reason within the parameters of Jewish law.

2. Stem cells for research purposes also can be procured from donated sperm and eggs mixed together and cultured in a petri dish. Genetic materials outside the uterus have no legal status in Jewish law, for they are not even a part of a human being until implanted in a woman's womb, and even then, during the first forty days of gestation, their status is "as if they were simply water."[5] Abortion is still prohibited during that time, except for therapeutic purposes, for in the uterus such gametes have the potential of growing into a human being. Outside the womb, however, at least at this time, they have no such potential. As a result, frozen embryos may be discarded or used for reasonable purposes and so may the stem cells that are procured from them.

OTHER FACTORS IN THIS DECISION

1. Given that the materials for stem cell research can be procured in permissible ways, the technology itself is morally neutral. It gains its moral valence on the basis of what we do with it.
2. The question, then, is reduced to a risk-benefit analysis of stem cell research. The articles in the most recent *Hastings Center Report*[6] raise some questions to be considered in such an analysis, but I will not rehearse them here. I want to note only two things about them from a Jewish perspective:

   a. The Jewish tradition views the provision of health care as a communal responsibility, and so the justice arguments in the *Hastings Center Report* have a special resonance for me as a Jew. Especially because much of the basic science in this area was funded by the government, the government has the right to require private companies to provide their applications of that science at reduced rates, or if necessary, at no cost, to those who cannot afford them. At the same time, the Jewish tradition does not demand socialism, and for many good reasons we in the United States have adopted a modified, capitalistic system of economics. The trick, then, will be to balance access to applications of the new technology with the legitimate right of a private company to make a profit on its efforts to develop and market those applications.

   b. The potential of stem cell research for creating organs for transplant and cures for diseases is, at least in theory, both awesome and hopeful. Indeed, in light of our divine mandate to seek to maintain life and health, one might even argue that from a Jewish perspective we have a duty to proceed with that research. As difficult as it may be, we must draw a clear line between uses of this or any other technology for cure, which are to be applauded, as opposed to uses of this technology for enhancement, which must be approached with extreme caution. Jews have been the brunt of campaigns of positive eugenics both here, in the United States, and in Nazi Germany,[7] and so we are especially sensitive to creating a model human being that is to be replicated through the kind of genetic engineering that stem cell applications will involve. Moreover, when Jews see a disabled human being, we are not to recoil from the disability or count our blessings for not being disabled in that way; rather, we are commanded to recite a blessing thanking God for making people different.[8] Thus, in light of the Jewish view that all human beings are created in the image of God, regardless of their levels of ability or disability, it is imperative from a Jewish perspective that the applications of stem cell research be used for cure and not for enhancement.

RECOMMENDATION

My recommendation is that we take the steps necessary to advance stem cell research and its applications in an effort to take advantage of its great potential for good. We should do so, however, in such a way that we provide access to its applications to all Americans who need them and at the same time prohibit the development of applications intended to make all human beings fit any particular model of human excellence. Through this technology, we should seek to cure diseases and to appreciate the variety of God's creatures.

NOTES

1. For more on these and other fundamental assumption of Jewish medical ethics, and for the Jewish sources that express these convictions, see E. N. Dorff, *Matters of Life and Death: A Jewish Approach to Modern Medical Ethics* (Philadelphia: Jewish Publication Society, 1998), chap. 2.

2. Gen. 2:15.

3. For more on the Jewish stance on abortion, together with the biblical and rabbinic sources that state that stance, see Dorff, *Matters of Life and Death*, pp. 128–33, and D.M. Feldman, *Birth Control in Jewish Law: Marital Relations, Contraception, and Abortion as Set Forth in the Classic Texts of Jewish Law* (New York: New York University Press, 1968), reprinted under the title *Marital Relations, Abortion, and Birth Control in Jewish Law* (New York: Schocken, 1973), chap. 14 and 15.

4. For classical sources on this, see Dorff, *Matters of Life and Death*, chap. 9.

5. *Babylonian Talmud*, Yevamot 69b. Rabbi Immanuel Jakobovits notes that "40 days" in talmudic terms may mean just under two months in our modern way of calculating gestation, since the rabbis counted from the time of the first missed menstrual flow while we count from the time of conception, approximately two weeks earlier. See I. Jakobovits *Jewish Medical Ethics: A Comparative and Historical Study of the Jewish Religious Attitude to Medicine and Practice* (New York: Bloch Publishing Company, 1959, 1975), p. 275.

6. Symposium, "Human Primordial Stem Cells," *Hastings Center Report* (March-April 1999): 30–48.

7. See S. J. Gould, *The Mismeasure of Man* (New York: W.W. Norton and Company, 1996) and G.J. Annas and M.A. Grodin, *The Nazi Doctors and the Nuremberg Code: Human Rights in Human Experimentation* (New York: Oxford University Press, 1992).

8. For a thorough discussion of this blessing and concept in Jewish tradition, see C. Astor, ". . . *Who Makes People Different:" Jewish Perspectives on the Disabled* (New York: United Synagogue of America, 1985).

20
Testimony of Abdulaziz Sachedina, Ph.D., University of Virginia

Islamic Perspectives in Research with Human Embryonic Stem Cells

Abdulaziz Sachedina for National Bioethics Advisory Commission

T hank you very much for inviting me to give an Islamic perspective on human stem cell research. I do not represent a particular school of thought ("church") in Islam; rather, I speak for the
Islamic tradition in general, which is a textual tradition. I have been
able to examine a number of primary and secondary sources that
have been produced by scholars representing different schools of
thought. Two major sects or schools of thought, the Sunni, who form
the majority in the Muslim community, and the Shi`ite, who form
the minority, do not represent an Orthodox/Reform divide; instead,
they are both "orthodox" in the sense that both base their arguments
on the same set of texts that are recognized as authoritative by all of
their scholars. And yet, it is important to keep in mind the plurality
of interpretations displayed by the "traditionalists" and "conservatives" on the one hand, and the "liberals" on the other.

The ethical-religious assessment of research uses of pluripotent
stem cells derived from human embryos in Islam can be inferen-
tially deduced from the rulings of the Shari`a, Islamic law, that deal
with fetal viability and the sanctity of the embryo in the classical and

modern juristic decisions. The Shari`a treats a second source of cells, those derived from fetal tissue following abortion, as analogically similar to cadaver donation for organ transplantation in order to save other lives, and hence, the use of cells from that source is permissible. For this presentation, I have researched three types of sources in Islamic tradition to assess the legal-moral status of the human embryo: commentaries on the Koranic verses that deal with embryology; works on Muslim traditions that speak about fetal viability; and juridical literature that treats the question of the legal-moral status of the human fetus (al-janin).

Historically, the debate about the embryo in Muslim juridical sources has been dominated by issues related to ascertaining the moral-legal status of the fetus. In addition, in order to provide a comprehensive picture representing the four major Sunni schools and one Shi`ite legal school, I have investigated diverse legal decisions made by their major scholars on the status of the human embryo and the related issue of abortion in order to infer religious guidelines for any research that involves the human embryo.

Let me repeat here, as I did when I testified to the Commission about Islamic ethical considerations in human cloning, that since the major breakthrough in scientific research on embryonic stem cells that occurred in November 1998, I have not come across any recent rulings in Islamic bioethics regarding the moral status of the blastocyst from which the stem cells are isolated.

The moral consideration and concern in Islam have been connected, however, with the fetus and its development to a particular point when it attains human personhood with full moral and legal status. Based on theological and ethical considerations derived from the Koranic passages that describe the embryonic journey to personhood developmentally and the rulings that treat ensoulment and personhood as occurring over time almost synonymously, it is correct to suggest that a majority of the Sunni and Shi`ite jurists will have little problem in endorsing ethically regulated research on the stem cells that promises potential therapeutic value, provided that the expected therapeutic benefits are not simply speculative.

The inception of embryonic life is an important moral and social question in the Muslim community. Anyone who has followed Muslim debate over this question notices that its answer has differed at different times and in proportion to the scientific information available to the jurists. Accordingly, each period of Islamic jurisprudence has come up with its ruling (*fatwa*), consistent with the findings of science and technology available at that time. The search for a satisfactory answer regarding when an embryo attains legal rights has continued to this day.

The life of a fetus inside the womb, according to the Koran, goes through several stages, which are described in a detailed and precise manner. In the chapter entitled "The Believers" (24), we read the following verses:

> We created (*khalaqna*) man of an extraction of clay, then We set him, a drop in a safe lodging, then We created of the drop a clot, then We created of the clot a tissue, then We created of the tissue bones, then we covered the bones in flesh; thereafter We produced it as another creature. So blessed be God, the Best of creators (*khaliqin*).[1]

In another place, the Koran specifically speaks about "breathing His own spirit" after God forms human beings:

> Human progeny He creates from a drop of sperm; He fashions his limbs and organs in perfect proportion and breathes into him from His own Spirit (*ruh*). And He gives you ears, eyes, and a heart. These bounties warrant your sincere gratitude, but little do you give thanks.[2]

> And your Lord said to the angels: "I am going to create human from clay. And when I have given him form and breathed into him of My life force (*ruh*), you must all show respect by bowing down before him."[3]

The commentators of the Koran, who were in most cases legal scholars, drew some important conclusions from this and other passages that describe the development of an embryo to a full human person. First, human creation is part of the divine will that determines the embryonic journey developmentally to a human creature. Second, it suggests that moral personhood is a process and achievement at the later stage in biological development of the embryo when God says: "*thereafter* We produced him as another creature." The adverb "thereafter" clarifies the stage at which a fetus attains personhood. Third, it raises questions in Islamic law of inheritance as well as punitive justice, where the rights and indemnity of the fetus are recognized as a person, whether the fetus should be accorded the status of a legal-moral person once it lodges in the uterus in the earlier stage. Fourth, as the subsequent juridical extrapolations bear out, the Koranic embryonic development allows for a possible distinction between a biological and moral person because of its silence over a particular point when the ensoulment occurs.

Earlier rulings on indemnity for homicide in the Shari`a were deduced on the premise that the life of a fetus began with the appreciation of its palpable movements inside the mother's womb, which occurs around the fourth month of pregnancy. In addition to the Koran, the following tradition on creation of human progeny provided the evidence for the concrete divide in pre- and postensoulment periods of pregnancy:

> Each one of you possesses his own formation within his mother's womb, first as a drop of matter for forty days, then as a blood clot for forty days, then as a blob for forty days, and then the angel is sent to breathe life into him.[4]

Ibn Hajar al-`Asqalani (d. 1449) commenting on the above tradition says:

> The first organ that develops in a fetus is the stomach because it needs to feed itself by means of it. Alimentation has precedence over all other functions for in the order of nature growth depends on nutrition. It does not need sensory perception or voluntary movement at this stage because it is like a plant. However, it is given sensation and volition when the soul (nafs) attaches itself to it.[5]

A majority of the Sunni and some Shi`ite scholars make a distinction between two stages in pregnancy divided by the end of the fourth month (120 days) when the ensoulment takes place. On the other hand, a majority of the Shi`ite and some Sunni jurists have exercised caution in making such a distinction because they regard the embryo in the pre-ensoulment stages as alive and its eradication as a sin. That is why Sunni jurists in general allow justifiable abortion within that period, while all schools agree that the sanctity of fetal life must be acknowledged after the fourth month.

The classical formulations based on the Koran and the Tradition provide no universally accepted definition of the term "embryo." Nor do these two foundational sources define the exact moment when a fetus becomes a moral-legal being. With the progress in the study of anatomy and in embryology, it is confirmed beyond any doubt that life begins inside the womb at the very moment of conception, right after fertilization and the production of a zygote. Consequently, from the earliest stage of its conception, an embryo is said to be a living creature with sanctity whose life must be protected against aggression. This opinion is held by Dr. Hassan Hathout, a physician by training, who was unable to be here today. This scientific information has turned into a legal-ethical dispute among Muslim jurists over the permissibility of abortion during the first trimester and the destruction of unused embryos, which would, according to this information, be regarded as living beings in the *in vitro* fertilization clinics. Some scholars have called for ignoring the sanctity of fetal life and permitting its termination at that early stage.

A tenable conclusion held by a number of prominent Sunni and Shi`ite scholars suggests that aggression against the human fetus is unlawful. Once it is established that the fetus is alive, the crime against it is regarded as a crime against a fully formed human being. According to these scholars, science and experience have unfolded new horizons that have left no room for doubt in determining signs of life from the moment of conception. Yet, as participants in the act of creating and curing with God, human beings can actively engage in furthering the overall good of humanity by intervening in the works of nature, including the early stages of embryonic development, to improve human health.

The question that still remains to be answered by Muslim jurists in the context of embryonic stem cell research is, When does the union of a sperm and an ovum entail sanctity and rights in the Shari`a? Most of modern Muslim opinions speak of a moment beyond the blastocyst stage when a fetus turns into a human being. Not every living organism in a uterus is entitled to the same degree of sanctity and honor as is a fetus at the turn of the first trimester.

The anatomical description of the fetus as it follows its course from conception to a full human person has been closely compared to the tradition about three periods of forty-day gestation to conclude that the growth of a well-defined form and evidence of voluntary movement mark the ensoulment. This opinion is based on a classical ruling given by a prominent Sunni jurist, Ibn al-Qayyim (d. 1350):

> Does an embryo move voluntarily or have sensation before the ensoulment? It is said that it grows and feeds like a plant. It does not have voluntary movement or alimentation. When ensoulment takes place voluntary movement and alimentation is added to it.[6]

Since there is no unified juridical-religious body representing the entire Muslim community globally, different countries have followed different classical interpretations of fetal viability. Thus, for instance, Saudi Arabia, might choose to follow Ibn Qayyim; while Egypt might follow Ibn Hajar al-`Asqalani. We need to keep in mind that the same plurality of the tradition is operative in North America when it comes to making ethical decisions on any of the controversial matters in medical ethics. Nevertheless, on the basis of all the evidence examined for this testimony, it is possible to propose the following as acceptable to all schools of thought in Islam:

1. The Koran and the Tradition regard perceivable human life as possible at the *later* stages of the biological development of the embryo.
2. The fetus is accorded the status of a legal person only at the later stages of its development, when perceptible form and voluntary movement are demonstrated. Hence, in earlier stages, such as when it lodges itself in the uterus and begins its journey to personhood, the embryo cannot be considered as possessing moral status.
3. The silence of the Koran over a criterion for moral status (i.e., when the ensoulment occurs) of the fetus allows the jurists to make a distinction between a biological and a moral person, placing the latter stage after, at least, the first trimester of pregnancy.

Finally, the Koran takes into account the problem of human arrogance, which takes the form of rejection of God's frequent reminders to humanity that God's immutable laws are dominant in nature and that human beings cannot willfully interfere to cause damage to others. "The will of God" in the Koran has often been interpreted as the processes of nature uninterfered with by human action. Hence, in Islam, research on stem cells made possible by biotechnical intervention in the early stages of life is regarded as an act of faith in the ultimate will of God as the Giver of all life, as long as such an intervention is undertaken with the purpose of improving human health.

NOTES

1. *Koran* 24: 12–14.
2. Ibid. 41: 9–10.
3. Ibid. 38: 72–73.
4. Sahih al-Bukhari and Sahih al-Muslim, *The Book of Destiny* (Qadar).
5. Ibn Hajar al-'Asqalani, *Fath al-Bari fi Sharh al-Sahih al-Bukhari, Kitab al-Qadar*, 11: 482.
6. Ibn al-Qayyim, *Al-Tibyan fi aqsam al-Qur'an* (Cairo, 1933), p. 255.

Part 5

Policy Issues

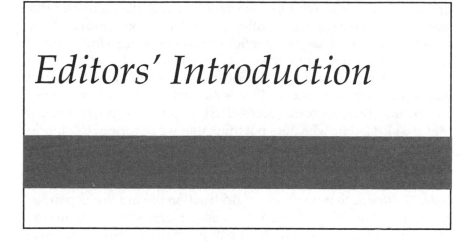

Editors' Introduction

W hat are we going to do? Remember, to do nothing is to do something. Frank E. Young does not want inaction, but he would like a moratorium on human embryo research until some of the ethical issues have been clarified. Kenneth Ryan, to the contrary, thinks we have spent far too much time dithering and doing nothing because special religious interests have blocked funding. The time has now come for action, and that means stem cell research and its application. The National Bioethics Advisory Commission in its report to President Clinton tried to walk a fine line between the extremes, arguing that government funding should not be allowed for work on embryos produced especially for stem cell research, but that it should be allowed for work on fetuses aborted for other reasons or on material discarded during fertility treatments. Probably this compromise will satisfy no one completely. Those in favor of research and its application will argue that in a society that already allows a million or more abortions a year for reasons that are often little more than convenience or cleaning up the effects of carelessness, it is hypocritical if naught else to forbid the abortion of fetuses produced especially with the aim of alleviating ill health and disease. Those against abortion will argue that abortion continues to be wrong, no matter what the intentions or consequences. But compromises are not intended to make everyone completely happy, rather to point a way forward, and it is clear that the commission's report had some effect on President Bush's August 9, 2001, ruling. Finally, we include two opposed pieces. First, a piece by neurobiologist Maureen L. Condic who does not think that we now need to support work on embryos with public funds. Second, a piece by philosopher Andrew W. Siegel that tries to uncover reasons behind a compromise, showing how all but extremists can allow some work on stem cells. "Research that involves the destruction of embryos is permissible where there is good reason to believe that it is necessary to cure life-threatening or severely debilitating diseases." Now you have enough material to see if this is a position with which you can agree.

21
A Time
for Restraint

Frank E. Young

The debate on the use of human embryos for research will be one of the more important issues of the twenty-first century. Unlike recombinant DNA technology, embryonic stem cell research most probably will result in the destruction of living embryos. Many people consider this research immoral, illegal, and unnecessary. Therefore, it is imperative to proceed cautiously. Federal funding of research using human embryos or pluripotent cells derived from them would be inappropriate until further resolution of the ethical issues has been achieved.

The ability to grow human embryonic stem (ES) cells *in vitro* challenges governments, regulatory agencies, and scientific organizations to define the ethical boundaries of using these cells in research. In the United States, President Clinton charged the National Bioethics Advisory Commission (NBAC) to review the medical and ethical considerations of this technology. In September 1999, the NBAC released the Executive Summary of its report.[1]

Reprinted with permission from *Science* 287, no. 5457 (February 2000): 1424. Copyright © 2000, American Association for the Advancement of Science.

While noting the existence of diverse views, it formulated a utilitarian approach, justifying public funding of research with human ES cells on the basis of the potential medical benefits. NBAC's primary concern was whether the "scientific merit and substantial clinical promise of this research justifies federal support, and if so with what restrictions and safeguards." Its ethical concern was focused on restricting the sources of embryos.

After National Institutes of Health director Harold Varmus publicly supported the use of human ES cells for research, based on a decision of the general counsel of the Department of Health and Human Services, seventy members of Congress signed a letter of objection. In a letter to *Science* in March 1999,[2] seventy-three scientists, including sixty-seven Nobel laureates, endorsed Varmus's position, claiming that it protects "the sanctity of life without impeding biomedical research." They noted many promising uses for ES cells, including therapeutic advances and reduction of animal studies and clinical trials needed for drug development. Again, the emphasis was on potential benefits, and the ethics of embryo destruction were not addressed.

The fact that experiments using ES cells, as currently performed, result in killing an embryo cannot be ignored so readily. In the United States, such action is in violation of many state laws that protect the embryo.[3] The disintegration of human embryos or the extraction of cells from blastocysts of human embryos [as has been described in primates][4] for the promised but as yet unrealized benefit of patients disregards concerns about the value of the individual that have already been raised by the prospect of human cloning.[5] The devaluation of humans at the very commencement of life encourages a policy of sacrificing the vulnerable that could ultimately put other humans at risk, such as those with disabilities and the aged, through a new eugenics or euthanasia.

Although there is great scientific interest in ES cell research, other recent advances suggest that adult stem cells may be more widely distributed than heretofore recognized and may thus obviate the need for ES cells.[6] Rather than risking public sanction and mistrust from those concerned with the ethical, legal, and moral status of the embryo, is it not wiser to give more than a passing mention to those concerns and in the meantime to do no harm to living embryos? It may be tempting to pursue a scientific imperative that impels us ever forward, but there are major costs. Regulatory policies and processes should take into account public confidence as well as the classical standards of safety and effectiveness. Our pluralistic society must consider the social, religious, medical, environmental, and scientific interests of its citizens. Once credibility is lost, acceptance is eroded.

Should scientific research be limited only by the value of its potential benefits? And who should make the decisions about the limits? To quote J. A. Robertson,[7] "Society, as the provider of the resources, the bearer of the costs, and the reaper of the benefits, has an overriding interest in the consequences of science, hence an interest in the routes and direction that research takes." Scientists who proclaim First Amendment freedom of inquiry are countered by a public suspicion of an inherent conflict of interest when their research support depends on funding from federal and industrial sources. Therefore, any commission regulating research should be composed of individuals of many persuasions and should include people who have no direct or indirect dependence on public monies.

Observations from the recombinant DNA (rDNA) debate can be useful in considering policies regarding human embryo research. The process of policy development was public, and committees consisted of individuals with diverse expertise, opinions, and backgrounds, including those who opposed rDNA research. Although public debate was contentious, careful analysis of the issues prevailed. The results demonstrated that scientists and the public can successfully work together to decide the appropriate use of public funds and formulate regulatory guidelines.[8]

The following recommendations are made to facilitate a consensus. Every nation conducting ES cell research should develop a national policy. In the United States, a representative commission should be appointed to review ES and adult stem cell research, develop an ethical framework for such research, and communicate with the public. It should also examine the adequacy of current guidelines and regulations for *in vitro* fertilization. A three-year moratorium on human embryo research should be instituted while the commission completes its work. Sufficient funding for research on human adult stem cells and animal embryonic and germinal stem cells should be provided during the moratorium.[9] International harmonization of guidelines could be accomplished through the Organization for Economic Cooperation and Development.

To rush to approve the destruction of embryos in order to harvest and experiment on ES cells is inadvisable and unnecessary. We should address the ethical concerns first.

NOTES

1. National Bioethics Advisory Commission (NBAC), *Ethical Issues in Human Stem Cell Research, Executive Summary* [online], http://bioethics.gov/cgi-bin/bioeth-counter.pl [September 1999].

2. R. P. Lanza et al., "Science over Politics," *Science* 283 (1999): 1849.

3. C. D. Forsythe, "Human Cloning and the Constitution," *Valparaiso University Law Review* 32 (1998): 491.

4. A. W. S. Chan, "Clonal Propagation of Primate Offspring by Embryo Splitting," *Science* 287 (2000): 317.

5. C. S. Campbell, "Religious Perspectives on Human Cloning" in *Cloning Human Beings*, vol. 2 of *Report and Recommendations of the National Bioethics Advisory Commission* (Rockville, Md.: NBAC, 1997), pp. D3–D60.

6. C. Frisenet al., "Central Nervous System Stem Cells in the Embryo and Adult," *Cellular and Molecular Life Sciences* 54 (1998): 935; C. B. Johansson et al., "Identification of a Neural Stem Cell in the Adult Mammalian Central Nervous System," *Cell* 96 (1999): 25; S. Temple and A. Alvarez-Buylla, "Stem Cells in the Adult Mammalian Central Nervous System," *Current Opinion in Neurobiology* 9 (1999): 135; H. G. Kuhn and C. N. Svendsen, "Origins, Functions, and Potential of Adult Neural Stem Cells," *BioEssays* 21 (1999): 625; M. Luquin et al., "Recovery of Chronic Parkinsonian Monkeys by Autotransplants of Carotid Body Cell Aggregates into Putamen," *Neuron* 22 (1999): 743.

7. J. A. Robertson, "Scientist's Right to Research: A Constitutional Analysis," *Southern California Law Review* 51 (1978): 1278.

8. "Coordinated Framework for Regulation of Biotechnology; Announcement of Policy and Notice for Public Comment," *Federal Register* 51 (1986): 23302.

9. Senate Appropriations Subcommittee on Labor, Health and Human Services, and Education, November 4, 1999, testimony, "Stem Cell Research," three speakers: Strom Thurmond (Senate), Jay Dickey (House of Representatives), and Dr. Frank Young (FDA).

22
The Politics and Ethics of Human Embryo and Stem Cell Research

Kenneth J. Ryan, M.D.

THE PROMISE OF HUMAN EMBRYO AND STEM CELL RESEARCH

Significant advances in cellular and developmental biology have occurred with animal research including the dramatic and successful cloning through adult somatic cell nuclear transfer for the sheep named Dolly in 1997.[1] Such cloning has now occurred in mice and cows[2] as well as sheep, and holds great promise in animal husbandry and for application to the manufacture of genetically designed drugs or proteins in sheep or cow milk for use in human medicine. For the time being, there is little interest and much anxiety in applying cloning to human reproduction and many ethical and safety questions remain to be considered. Much has been accomplished otherwise in human reproduction since 1978, after the first human birth with *in vitro* fertilization of Louise Brown.[3] With

Reprinted from Kenneth J. Ryan, "The Politics and Ethics of Human Embryo and Stem Cell Research," *Women's Health Issues* 10, no. 3 (May/June 2000): 105–10.

this information as background, the promise of future advances using human embryo research seems bright and includes better diagnosis and treatment for male and female infertility, strategies for preventing congenital defects such as preimplantation genetic diagnosis, as well as studies on early human development, fertilization, and implantation. It is, however, the isolation of human stem cells from the inner cell mass of embryos or primordial germ cells of fetuses and the potential for their differentiation in tissue culture into various other cell types such as nerve or blood or heart cells for use in treatment of disease that has captured the imagination of the public and excited scientists. This also might ultimately decrease dependence on organ transplants, for which there has been a chronic shortage and disputes over regional allocations. The isolation of human stem cells from embryos and fetal primordial germ cells occurred in 1998,[4] and its great promise immediately prompted an ethical discussion about the rules then in place forbidding federal funding for research involving human embryos. Applications for use of this stem cell technique have been proposed for such problems as Parkinson's and Alzheimer's diseases, diabetes, heart disease, spinal cord injury, and metabolic diseases in young children. Furthermore, if the cells for transplanting were derived from the insertion of a patient's somatic cell nucleus into an egg or possibly a pluripotent stem cell, the resulting clonal cell line would be expected to have no tissue incompatibility with the donor of the cell nucleus. This would avoid tissue rejection and the need for chronic use of antirejection drugs. Taking this use of stem cells out of the realm of science fiction, scientists already reported in 1999 on the conversion of stem cells from a mouse embryo into specialized brain cells that corrected a myelin deficiency when injected into mice genetically incapable of synthesizing myelin.[5] Beyond theoretical therapeutic applications, there is also the potential for advances in basic knowledge of normal and cancer cell development that can create new opportunities for application to human medicine. The great promise, broad application, and wide support of scientists for this research are important factors in developing public policy about the role of government in the funding, regulation, and ultimate application of such research despite ethical concerns about the use of human embryos or aborted fetuses as the source for research material.

HISTORY OF REGULATIONS GOVERNING FEDERALLY FUNDED RESEARCH ON FETUSES AND EMBRYOS

After the *Roe* v. *Wade* supreme court decision in 1973, there was increased concern in the U.S. Congress that live fetuses in pregnancies destined for abortion would be exploited for research purposes and that abortions might be encouraged largely to satisfy research needs. Before this, there had been reports about abusive research funded by the National Institutes of Health on midtrimester live fetuses both in this country and overseas. As a consequence, a moratorium on federal funding of fetal research was imposed until a national commission created by the National Research Act of 1974 could consider how such abuses could be prevented. The commission report in 1975 provided strict regulatory guidelines for fetal research and the moratorium was lifted.[6] At this time, support for research on tissue derived from deceased aborted fetuses was acceptable. No decision was made on the then prohibited federal funding of research on *in vitro* fertilization and preimplantation embryos because this was to be reviewed by an ethics advisory board to be appointed by the Secretary of Health, Education, and Welfare. The Ethics Advisory Board was formed two years later in 1977 and issued a report in 1978 after the first *in vitro* fertilization baby was born in England. The Ethics Advisory Board concluded that research on very early embryos within the first fourteen days of development was acceptable to develop techniques for *in vitro* fertilization if the gametes were derived from a married couple.[7] (The requirement in 1978 for a couple to be married to provide gametes for pregnancy seems a quaint touch in the present era.) Funding for such research on the early embryo was blocked by pro-life members of Congress and the administrations of both Presidents Ronald Reagan and George Bush, who objected to the inevitable loss of early embryos in the process of developing safe techniques. The favorable report was never implemented and the Ethics Advisory Board was disbanded, resulting in a de facto moratorium on embryo research until a Democratic Congress and president were elected in 1992.

In 1987 and 1988, research on the use of fetal tissue from deceased aborted fetuses for transplantation into patients suffering from Parkinson's disease became an issue and a moratorium was imposed on federal funding until an advisory panel to the director of the National Institutes of Health (NIH) could consider the matter. The chief ethical concerns were whether this would encourage abortion and whether use of such tissue derived from aborted fetuses was unethical because of possible complicity with the abortion. The advisory panel voted nineteen-to-two that with safeguards, federal funding for research on fetal use was appropriate.[8] Nevertheless, proposed legislation to permit such funding was vetoed by President Bush and funding for the use of fetal tissue did not resume until a new Congress and administration were elected and changed the regulations in 1993. Also in 1993, the National Institutes of Health Revitalization Act rescinded the requirement for an ethics advisory board to review research involving preimplantation embryos. A Human Embryo Research Panel was appointed by the director of the NIH to consider whether research on embryos should be conducted, and if so, what review and guidelines were needed for the research to proceed. There was a request for three categories: research acceptable for federal funding, research needing more review before a decision could be made, and research unacceptable for federal funding. The panel recommendation allowed, for example, the use of spare embryos not needed after an *in vit*ro fertilization cycle when consent from gamete donors had been obtained. The permissible research categories included the development of embryonic stem cells as well as work on fertilization, improving pregnancy outcomes, and freezing of oocytes. A type of research not considered acceptable was human cloning for purposes of reproduction. Needing more review, for example, was research on stem cell development from embryos created for that purpose.[9] President Clinton quickly determined that embryos should not be created specifically for research purposes and immediately prohibited funding for this category of research. In the end, funding of any research found acceptable by the panel has not been possible because Congress has regularly added a rider to the annual budget of the NIH prohibiting any research that results in the destruction of an embryo. A strong stand on permitting embryo research to proceed by a broad consensus of the scientific community did not occur until the promise of stem cell research became appreciated. The NIH is currently considering getting around the congressional ban on embryo research by funding research on human stem cells after they are formed, but not funding the actual isolation of the stem cells from embryos which would be accomplished in the private sector.[10] Congressmen are not pleased with this political maneuver, which they consider dealing with the letter and not the intent of their legislation. It is interesting that objections to embryo research could not be overcome for study of problems related to women's reproductive interests, but required the more general promise of the exciting use of stem cells to forge a political will to fight for the research.

In fact, there has been a moratorium on federal funding of embryo research for twenty years, and consequently all improvements in assisted reproduction have to be derived from research funded by the private sector. This has created a sort of moral vacuum with no federal oversight and few research guidelines. During this period, women's interests in advances in reproductive science have been hostage to a conservative agenda in government. Even dues owed by the United States to the United Nations have been withheld for years by a conservative pro-choice Congress. It will be interesting to see who prevails in the standoff over stem cell research: the NIH backed by a broad science consensus or a coalition of seventy pro-life congressmen. Federal support for research involving aborted fetuses finally was regained when the promise of tissue transplantation for Parkinson's and other diseases became great enough to marshal a political will for change.

ETHICAL CONSIDERATIONS OF HUMAN EMBRYO AND STEM CELL RESEARCH

Public discourse on the ethics of abortion started in earnest and became mean-spirited after the Supreme Court 1973 *Roe v. Wade* decision. What had been a shadowy private and often illegal matter came into full public view and entered the political mainstream. The issue of research on aborted fetuses became a prominent part of the congressional concern in 1973 lest it increase the number of abortions or contribute to a sense of good derived from an immoral act.

The ethical arguments are by now familiar. One formula is that life is sacred and begins at conception. Therefore, abortion is always wrong unless perhaps the pregnant woman's life is in danger, although some pro-life activists deny an exemption even for threats to the woman's life. The counterargument states that the need for respect of fetal life is not absolute. At fertilization the beginning life starts with a single cell without a nervous system, feelings, interests, or rights, and the pregnant woman's needs and wants should be controlling. During the first trimester, the primitive status of the fetus progressively changes. As the fetus develops, its relative standing should change, and after viability it starts assuming status in society until at birth it assumes full protection under the law. Between these two extremes are many formulations including the familiar and more broadly accepted justifications for abortion. Examples are that the pregnancy threatens a woman's life or health, including mental health, and reasons including rape, incest, and severe fetal defect. The right to abortion in the first trimester seems widely accepted. In addition, the conventional wisdom is that early abortions are preferable on all counts, but after viability the justification for an abortion should be more stringent. All of those arguments have been challenged on the basis of whether they are religious in nature and therefore are not controlling in a pluralistic secular society. In addition, all sorts of metaphors and rhetoric have been employed to sway public opinion, such as whether the right and wrong of abortion are similar to slavery in America before Lincoln emancipated the slaves, or like the Nazi holocaust; this method of persuasion further inflames the passions in the dispute.[11] Although the rhetoric may seem superficially appealing, most of the analogies simply to do not stand up to careful scrutiny. It has also been denied that there are many true cases of rape or incest and that women lie about this to obtain an abortion. This attack on the rape victim's claim seems part of a cruel and pervasive history of sex discrimination in a male-dominated world in the face of the public record on abuse of women. The idea that women ought to be responsible for their sexual acts and should pay the price if they get pregnant is also part of the old repressive puritanical response for what is presumed immoral behavior.[12]

The use of an abortion argument to prevent research on the dead fetus or the use of tissue for treating problems such as Parkinson's disease is part of an agenda to keep abortion a political issue. The use of tissue for these purposes from spontaneously aborted fetuses, or from deceased children and adults, although less often useful, has not engendered the same fierce argument. On the debate over whether one can use fetal tissue or conduct research after an induced abortion, the argument has been that the use of tissue from aborted fetuses will increase the numbers of abortion or justify them, and that in any case the researchers become complicit in an immoral act. Finally, it has been claimed that a woman seeking an abortion should not have the right to give consent to the disposition of the fetus; the counterargument is that although the rationale of this claim is denied, it is not clear who else has standing to give or withhold consent.

Arguments about the ethics of research on the preimplantation embryo and the work on stem cells derived from embryos or fetuses follow those used against abortion, except the conflict between a pregnant woman's rights and status of the fetus she carries are not relevant. The argument here is about the moral status of the developing embryo in the laboratory, usually in the first week or two of development, where the issue of sacredness and rights and interests are raised. Again at issue is the question of the absolute sanctity of all life even at its inception versus a symbolic regard for life that grows with its development. Some, including President Clinton, could not justify creating an embryo *in vitro* just for research; hence, his directive forbidding federal support for this. However, most people believe that embryos created for infertility that are no longer needed and otherwise would be discarded could be used for research to advance knowledge in reproduction and benefit humankind generally. The offshoot of federal support for stem cell development from embryos seems a reasonable extension of this belief. The complicity argument of automatic responsibility for the loss of an embryo destined for discard anyway seems as far-fetched as complicity for abortion of a fetus with the subsequent use of its tissue when the events are kept separate by law and regulation.[13] It should be obvious that assisted reproduction technologies may produce excess embryos in infertility cycles which can be frozen for later use, but discard of such excess embryos ultimately will be a problem whether or not they are used for research.

None of the ethical arguments seem destined to change made-up minds on the fundamentals of how sacred and absolute life in the womb or test tube is at conception and immediately thereafter. Claims on behalf of the embryo compete with a woman's right to control her reproductive interests, and with society's pursuit of beneficial research in the areas of reproduction and general biology and medicine. Arguments based on principle on the status of life at its beginning have not changed many minds and are unlikely to in the future. There is only an appeal to a point of view and resulting convictions. In such a case, common sense and the consequences of a course of action should be controlling. We never have and are unlikely ever to grant full citizenship and personhood to the fertilized egg or contingent life before viability. The benefits from carefully regulated research are undeniably enormous. Arguments that this is a cause of abortions or a complicity in immoral acts are not convincing. In my view, public policy should favor governmental support for embryo and stem cell research.

NOTES

1. I. Wilmut et al., "Viable Offspring Derived from Fetal and Adult Mammalian Cells," *Nature* 385 (1998): 810–13.

2. T. Wakayama et al., "Full-Term Development of Mice from Enucleated Oocytes Injected with Cumulus Cell Nuclei," *Nature* 394 (1998): 369–74; Y. Kato et al., "Eight Calves Cloned from Somatic Cells of a Single Adult," *Science* 282 (1998): 2095–98.

3. P. C. Steptoe and R. G. Edwards, "Birth after Reimplantation of a Human Embryo," *Lancet* 2 (1978): 366.

4. J. A. Thomson et al., "Embryonic Stem Cell Lines Derived from Human Blastocysts," *Science* 282 (1998): 1145–47; M. J. Shamblott et al., "Derivation of Pluripotent Stem Cells from Cultured Human Primordial Germ Cells," *Proceedings of the National Academy of Sciences USA* 95 (1998): 13726–31.

5. N. Wade, "Mouse Cells Are Converted into Brain Cells," *New York Times,* July 30, 1999, A13.

6. National Commission for the Protection of Human Subjects of Biomedical and Behavioral Research, *Report and Recommendations: Research on the Fetus* (Washington, D.C.: Department of Health Education and Welfare, 1975).

7. J. C. Fletcher and K. J. Ryan, "Federal Regulations for Fetal Research: A Case for Reform" in *Fetal Diagnosis and Therapy,* ed. M. I. Evans et al., (Philadelphia: J. B. Lippincott, 1989).

8. National Institutes of Health (NIH), *Report of the Panel on Fetal Tissue Transplantation Research* (Bethesda, Md.: NIH, 1989).

9. NIH Human Embryo Research Panel, *Report of the Human Embryo Research Panel*, 2 vols. (Bethesda, Md.: NIH, 1994).

10. N. Wade, "Advisory Panel Votes for Use of Embryonic Cells in Research," *New York Times*, June 29, 1999, A13.

11. C. M. Condit, *Decoding Abortion Rhetoric* (Urbana: University of Illinois Press, 1990).

12. K. J. Ryan, "Abortion or Motherhood, Madness and Suicide," *American Journal of Obstetrics and Gynecology* 166 (1992): 1029–36.

13. J. A. Robertson, "Ethics and Policy in Embryonic Stem Cell Research," *Kennedy Institute of Ethics Journal* 9 (1999): 109–36.

23
Ethical Issues in Human Stem Cell Research

Conclusions and Recommendations

National Bioethics Advisory Commission

INTRODUCTION

In November 1998, President Clinton charged the National Bioethics Advisory Commission with the task of conducting a thorough review of the issues associated with human stem cell research, balancing all ethical and medical considerations. The president's request was made in response to three separate reports that brought to the fore the exciting scientific and clinical prospects of stem cell research while also raising a series of ethical controversies regarding federal sponsorship of scientific inquiry in this area. Such research raises ethical issues because it involves the derivation of human *embryonic germ (EG) cells* from aborted fetuses or the derivation of human *embryonic stem (ES) cells* from early-stage embryos remaining after infertility treatments. A number of these important ethical concerns previously have been identified in public debate, both here and abroad. The commission reviewed these concerns in

Reprinted with permission from the National Bioethics Advisory Commission (NBAC).

light of both the medical and scientific promise in this significant new field and the existing statutes and regulations that affect research in this area. Our task, however, was neither to engage in moral analysis for its own sake nor to address all the regulatory issues that might be raised, but rather to offer advice on how the balance of ethical, scientific, and medical considerations should shape policies on the use of federal funds to support research that involves deriving or using human ES or EG cells.

SCIENTIFIC AND MEDICAL CONSIDERATIONS

The stem cell is a unique and essential cell type found in animals. Many kinds of stem cells are found in the human body, with some more differentiated, or committed, to a particular function than others. In other words, when stem cells divide, some of the progeny mature into cells of a specific type (heart, muscle, blood, or brain cells), while others remain stem cells, ready to repair some of the everyday wear and tear undergone by our bodies. These stem cells are capable of continually reproducing themselves and serve to renew tissue throughout an individual's life. For example, they continually regenerate the lining of the gut, revitalize skin, and produce a whole range of blood cells. Although the term *stem cell* commonly is used to refer to those cells within the adult organism that renew tissue (e.g., hematopoietic stem cells, a type of cell found in the blood), the most fundamental and extraordinary of the stem cells are found in the early-stage embryo. These ES cells, unlike the more differentiated adult stem (AS) cells or other cell types, retain the special ability to develop into nearly any cell type. EG cells, which originate from the primordial reproductive cells of the developing fetus, have properties similar to ES cells.

It is the potentially unique versatility of the ES and EG cells derived, respectively, from the early stage embryo and cadaveric fetal tissue that presents such unusual scientific and therapeutic promise. Indeed, scientists have long recognized the possibility of using such cells to generate more specialized cells or tissue, which could allow the newly generated cells to be used to treat injuries or diseases such as Alzheimer's disease, Parkinson's disease, heart disease, and kidney failure. In addition, scientists regard these cells as important—perhaps essential—in understanding the earliest stages of human development and in developing life-saving drugs and cell-replacement therapies to treat disorders caused by early cell death or impairment. At the same time, the techniques for deriving these cells have not been fully developed as standardized and readily available research tools, and the development of any therapeutic application remains some years away.

Research also is under way to determine whether human stem cells could be obtained from the differentiated stem cells of fully developed organisms. Thus far, however, studies in animals indicate that this approach faces substantial scientific and technical limitations; indeed, the anatomic source of certain cells might preclude easy or safe access in human beings. In addition, important biological differences apparently exist between ES cells, EG cells, and AS cells. Furthermore, differences among species mean that for full scientific and clinical benefits to be realized, some research will need to be conducted with human ES and EG cells, even as the emphasis remains on laboratory and animal research. In summary, research using stem cells from animals or from human adults is not a substitute for human ES and EG cell research, and it is toward the latter that we direct our ethical and policy analyses.

ETHICAL AND POLICY CONSIDERATIONS

The longstanding controversy about the ethics of research involving human embryos and cadaveric fetal material arises from fundamental and sharply differing moral views regarding elective abortion or the use of embryos for research. Indeed, an earnest national and international debate continues over the ethical, legal, and medical issues that arise in this arena. This debate represents both a challenge and an opportunity: a challenge because it concerns important and morally contested questions regarding the beginning of life, and an opportunity because it provides another occasion for serious public discussion about important ethical issues. We are hopeful that this dialogue will foster public understanding about the relationships between the opportunities that biomedical science offers to improve human welfare and the limits set by important ethical obligations.

Although we believe most would agree that human embryos deserve respect as a form of human life, disagreements arise regarding both what form such respect should take and what level of protection is required at different stages of embryonic development. Therefore, embryo research that is not therapeutic to the embryo is bound to raise serious concerns for some about how to resolve the tensions between two important ethical commitments: to cure disease and to protect human life. For those who believe that from the moment of conception the embryo has the moral status of a person, research (or any other activity) that would destroy the embryo is considered wrong and should be prohibited. For those who believe otherwise, arriving at an ethically acceptable policy in this arena involves a complex balancing of many important ethical concerns. Although many of the issues remain contested on moral grounds, they can exist within a broad area of consensus upon which public policy can, at least in part, be constructed.

For most observers, the resolution of these ethical and scientific issues depends to some degree upon the source of the stem cells. The use of cadaveric fetal tissue to derive EG cell lines—like other uses of tissues or organs from dead bodies—is generally the most acceptable of these sources, provided that the research complies with the system of public safeguards and oversight already in place for such scientific inquiry. With respect to embryos and the ES cells from which they can be derived, some draw an ethical distinction among three potential types of embryos. One is referred to as the *research embryo*, an embryo created through *in vitro* fertilization (IVF), with gametes provided solely for research purposes. Many people, including the president, have expressed the view that the federal government should not fund research that involves creating such embryos. The second type of embryo is that which was created for treatment of infertility, but is now intended to be discarded because it is unsuitable or no longer needed for such treatment. The use of these embryos raises fewer ethical questions because it does not alter their final disposition. Finally, the recent demonstration of cloning techniques (somatic cell nuclear transfer [SCNT]) in non-human animals suggests that the transfer of a human somatic cell nucleus into an oocyte might create an embryo that could be used as a source of ES cells. The creation of a human organism using this technique raises questions similar to those raised by the creation of research embryos through IVF, and at this time federal funds may not be used for such research. In addition, if the enucleated oocyte that was to be combined with a human somatic cell nucleus came from a nonhuman animal, other issues would arise about the nature of the embryo produced. Thus, each source of material raises distinct ethical questions as well as scientific, medical, and legal ones.

Conscientious individuals have reached different conclusions regarding both public policy and private actions in the area of stem cell research. Their differing perspectives by their very nature cannot easily be bridged by any single public policy. But the development of such policy in a morally contested area is not a novel challenge for a pluralistic democracy such as that which exists in the United States. We are profoundly aware of the diverse and strongly held views on the subject of this report and have wrestled with the implications of these different views at each of our meetings devoted to this topic. Our aim throughout these deliberations has been to formulate a set of recommendations that fully reflects widely shared views and that, in our view, would serve the best interests of society.

Most states place no legal restrictions on any of the means of creating ES and EG cells that are described in this report. In addition, current Food and Drug Administration (FDA) regulations do not apply to this type of early stage research. Therefore, because the public controversy surrounding such activities in the United States has revolved around whether it is appropriate for the federal government to sponsor such research, this report focuses on the question of whether the scientific merit and the substantive clinical promise of this research justify federal support, and, if so, with what restrictions and safeguards.

Views about the status of embryos and fetuses vary widely. Some believe that what matters is the potential for a new human life that arises at the moment of conception, while others identify the relevant concept as personhood, which they say begins only at some postembryonic stage. We heard from many members of the public, including those who are eager for this area of research to move forward as rapidly as possible, as well as those who oppose the research if it is built upon any activity that is connected to abortion or to the destruction of fertilized human eggs. In addition, our deliberations have been informed by testimony from scientists and physicians, lawyers and other experts in governmental regulation, philosophers, and Catholic, Protestant, Jewish, Islamic, and Eastern Orthodox theologians. As a result of these discussions, it has become clear that the question of whether federal policy should sponsor human ES or EG cell research is characterized by a tension between the desire to realize the great therapeutic benefits that may be derived from such work and the need to recognize that the materials involved must be treated with respect. We concluded that sufficient safeguards can be put in place in order to prevent abuse and to ensure that any use of embryos that remain after infertility treatments—like any use of fetal remains following elective abortion—is based upon and embodies the kind of respect for the embryos that most Americans would expect and demand of any activity that is carried out with the support of the federal government. Beyond the regulatory effects of the rules adopted to govern federal support for research in this area—with which we hope private sponsors of research involving ES and EG cells will comply voluntarily—the states also can influence research in this field through statutes and regulations on abortion, embryo research, and the donation of human body parts, embryos, and gametes.

CONCLUSIONS AND RECOMMENDATIONS

The conclusions and recommendations presented in this chapter are grouped into several categories:

- the ethical acceptability of federal funding for research that either derives or uses ES or EG cells
- the requirements for the donation of cadaveric fetal tissue and embryos for research
- restrictions on the sale of these materials and designation of those who may benefit from their use
- the need—and the means—for national oversight and institutional review
- the need for local review of derivation protocol
- the responsibilities of federal research agencies
- the issues that must be considered regarding the private sector
- the need for ongoing review and assessment

The Ethical Acceptability of Federal Funding of ES Cell and EG Cell Research

Despite the enormous scientific and clinical potential offered by research use of ES or EG cells derived from various sources, many find that certain sources are more ethically problematic than others. Our recommendations reflect respect for these diverse views, which varied even among the commissioners, regarding the ethical acceptability of the derivation and use of ES and EG cells from various sources. Human ES and EG cells can be derived from the following sources:

- human fetal tissue following elective abortion (EG cells)
- human embryos that are created by IVF and that are no longer needed by couples being treated for infertility (ES cells)
- human embryos that are created by IVF with gametes donated for the sole purpose of providing research material (ES cells)
- potentially, human (or hybrid) embryos generated asexually by SCNT or similar cloning techniques in which the nucleus of an adult human cell is introduced into an enucleated human or animal ovum (ES cells)

A principal ethical justification for public sponsorship of research with human ES or EG cells is that this research has the potential to produce health benefits for those who are suffering from serious and often fatal diseases. We recognize that it is possible that all the various sources of human ES or EG cells eventually could be important to research and clinical application because of, for example, their differing proliferation potential, differing availability and accessibility, and differing ability to be manipulated, as well as possibly significant differences in their cell biology.

Although each source of stem cells poses its own scientific, ethical, and legal challenges and opportunities, much of the ethical analysis leading to public policy recommendations depends upon the scientific and/or clinical necessity of using a specific source of the cells. In our judgment, the immediate scientific uses of ES or EG cells can be satisfied by the derivation and use of cell lines derived from fetal tissues (i.e., EG cells) and from embryos (i.e., ES cells) remaining after infertility treatments have ended.

The potential use of matched tissue for autologous cell-replacement therapy from ES cells may in the future require the use of cell lines developed by SCNT techniques. In addition, embryos created through IVF specifically as a source of ES cells might be essential for creating banks of multiple cell lines representing a spectrum of alleles for the major histocompatibility complex. This goal might require that ova and sperm of persons with specific genotypes be selected to make embryos from which to derive particular classes of ES cells.

Finally, although much promising research currently is being conducted with stem cells obtained from adult organisms, studies in animals suggest that this approach will be scientifically and technically limited, and, in some cases, the anatomic source of the cells might preclude easy or safe access. Important research can and should go forward in this area, although important biological differences exist between ES and AS cells, and the use of AS cells should not be considered an alternative to ES and EG cell research.

Much research into the generation of specific tissue types from stem cells can be conducted using EG cells derived from fetal material and ES cells derived from embryos remaining after infertility treatments. In the future, adequate scientific evidence and increased prospect for medical benefits may be available to generate public support for using human ES cells derived from embryos produced through IVF for research purposes or by SCNT for autologous transplant. We note, however, that a responsible federal science policy does not necessarily require public funding for access to all sources of ES or EG cells at once. At this time, therefore, the commission believes that federal funding for the use and derivation of ES and EG cells should be limited to two sources of such material: cadaveric fetal tissue and embryos remaining after infertility treatments. Specific recommendations and their justifications are provided below.

Recommendation 1

Research involving the derivation and use of human EG cells from cadaveric fetal tissue should continue to be eligible for federal funding. Relevant statutes and regulations should be amended to make clear that the ethical safeguards that exist for fetal tissue transplantation also apply to the derivation and use of human EG cells for research purposes.

Considerable agreement exists, both in the United States and throughout the world, that the use of fetal tissue in therapy for those with serious disorders, such as Parkinson's disease, is acceptable.[1] Research that uses cadaveric tissue from aborted fetuses is analogous to the use of fetal tissue in transplantation. The rationales for conducting EG research are equally strong, and the arguments against it are not persuasive. The removal of fetal germ cells does not occasion the destruction of a live fetus, nor is fetal tissue intentionally or purposefully created for human stem cell research. Although abortion itself doubtless will remain a contentious issue in our society, the procedures that have been developed to prevent fetal tissue donation for therapeutic transplantation from influencing the abortion decision offer a model for creating such separation in research to derive human EG cells. Because the existing statutes are written in terms of tissue transplantation, which is not a current feature of EG cell research, changes are needed to make explicit that the relevant safeguards will apply to research to derive EG cells from aborted fetuses.

Due to the contentious and polarizing nature of the abortion debate in the United States, restrictions were enacted over a decade ago to block the use of federal funding of fetal tissue transplantation therapy research. Until 1993, the only permissible source of tissue for such research was tissue from spontaneously aborted fetuses or ectopic pregnancies—sources that were largely unsuitable for research. In 1993, President Clinton lifted the ban on the use of fetal tissue from elective abortions for fetal tissue transplantation research.

Previous moral opposition to fetal tissue transplant research, because of its association with abortion, helped shape a system of safeguards to prevent the encouragement of the practice. These rules require that the consent process for women making abortion decisions must precede separately from the consent process for donation of fetal tissue for transplant research. Although some disagree, sufficient consensus exists that society should respect the autonomous choices of women who have chosen to have legal abortions to donate fetal tissue for research. If women have a right to choose to have an abortion, it follows that the self-determination or autonomy expressed in that right extends to the choice to donate fetal tissue for research purposes. Research using fetal tissue obtained after legal elective abortions will greatly benefit biomedical science and also will provide enormous therapeutic benefits to those suffering from various diseases and other conditions. In our view, there is no overriding reason for society to discourage or prohibit this research and thus forgo an important opportunity to benefit science and those who are suffering from illness and disease—especially in light of the legality of elective abortions that provide access to fetal tissue and of the risks involved in losing these valuable opportunities. Indeed, the consequences of forgoing the benefits of the use of fetal tissue may well be harmful. Moreover, if not used in research, this tissue will be discarded.

The Acceptability of Federal Support for Research Using Embryos Remaining After Infertility Treatments to Derive ES Cells

The current congressional ban on embryo research prohibits federal support of any research "in which a human embryo ... [is] destroyed, discarded, or knowingly subjected to risk of injury greater than that allowed for research on fetuses *in utero*."[2] The term "human embryo" in the statute is defined as "any organism, that is derived by fertilization, parthenogenesis, cloning, or any other means from one or more human gametes or human diploid cells."

The ban, which concerns only federally sponsored research, reflects a moral stance either that embryos deserve some measure of protection from society because of their moral status as persons, or that sufficient public controversy exists such that federal funds should not be used for this type of research. However, some effects of the embryo research ban raise serious moral and public policy concerns for those who hold differing views of the ethics of embryo research. In our view, the ban conflicts with several of the ethical goals of medicine, especially healing, prevention, and research—goals that are rightly characterized by the principles of beneficence and nonmaleficence, jointly encouraging the pursuit of each social benefit and avoiding or ameliorating potential harm.

In the United States, moral disputes—especially those concerning certain practices in the area of human reproduction—are sometimes resolved by denying federal funding for those practices (e.g., elective abortion), while not interfering with the practice in the private sector. In this case, investigative embryo research guided only by self-regulation is a widespread practice in the private sector, and the ban on embryo research has served to discourage the development of a coherent public policy, not only regarding embryo research but also regarding health research more generally. The ban also may have more profound effects on other areas of federally supported research that are dedicated to the relief of human suffering, raising concerns about the distribution and allocation of federal research resources. For example, by limiting the federal government's ability to fund promising areas of basic research, a complete ban on embryo research could prevent promising, collaborative studies in other areas, such as cancer and genetics. We recognize that many factors affect how federal research priorities are set in this country. However, in our view, the intentional withholding of federal funds for research that may lead to promising treatments may be considered unjust or unfair.

Although no consensus has been reached regarding the moral status of the embryo, there is agreement that if embryo research is permissible, some limitations and/or regulations are necessary and appropriate. Such regulation reflects an appreciation of the disparate views regarding the acceptability and unacceptability of this area of scientific investigation and serves as a way of providing accountability, allaying public anxiety, promoting beneficial research, and demonstrating respect for human embryos.

Recommendation 2

Research involving the derivation and use of human ES cells from embryos remaining after infertility treatments should be eligible for federal funding. An exception should be made to the present statutory ban on federal funding of embryo research to permit federal agencies to fund research involving the derivation of human ES cells from this source under appropriate regulations that include public oversight and review. (See Recommendations 5 through 9.)

Based on advice from the Department of Health and Human Services (DHHS) General Counsel, the director of the National Institutes of Health (NIH) announced in January 1999 that NIH will apply the ban only to research involving the derivation of ES cells from human embryos but not to research involving only the use of

ES cells. NIH has indicated that research proposals that involve the use of ES cells will be considered for funding once NIH has established a set of special guidelines that are currently under development. The DHHS General Counsel concluded that ES cells are not, in themselves, organisms and hence cannot be embryos as defined by the statute. Thus, it could be surmised from this interpretation that the only activity that could amount to "research in which a human embryo or embryos are destroyed" would be an attempt to derive ES cells from living embryos. This, in fact, is the interpretation adopted by DHHS and NIH. More than 70 members of Congress have protested this interpretation, claiming that whatever the language of the statute, Congress clearly intended to prohibit not just research in which human embryos are destroyed but also research that depends on the prior destruction of a human embryo. Yet the plain meaning of the statutory wording differs from this interpretation, and nothing in its legislative history indicates that either proponents or opponents of the rider anticipated a situation in which research that destroyed the embryo would be conducted separately from research that used the cells derived from the embryo. Thus, in legal terms, the general counsel's interpretation appears to be reasonable, even though it does not address any of the ethical concerns involved.

Although some may view the derivation and use of ES cells as ethically distinct activities, we believe that it is ethically acceptable for the federal government to finance research that both derives cell lines from embryos remaining after infertility treatments and that uses those cell lines. Although one might argue that some important research could proceed in the absence of federal funding for research that derives stem cells from embryos remaining after infertility treatments (i.e., federally funded scientists merely using cells derived with private funds), we believe that it is important that federal funding be made available for protocols that also derive such cells. Relying on cell lines that might be derived exclusively by a subset of privately funded researchers who are interested in this area could severely limit scientific and clinical progress.

An ethical problem is presented in trying to separate research in which human ES cells are used from the process of deriving those cells, because doing so diminishes the scientific value of the activities receiving federal support. This division—under which neither biomedical researchers at NIH nor scientists at universities and other research institutions who rely on federal support could participate in some aspects of this research—rests on the mistaken notion that derivation and use can be neatly separated without affecting the expansion of scientific knowledge. We believe that this misrepresentation of the new field of human stem cell research has several implications.

First, researchers using human ES cell lines will derive substantial scientific benefits from a detailed understanding of the process of ES cell derivation, because the properties of ES cells and the methods for sustaining the cell lines may differ depending upon the conditions and methods used to derive them. Thus, scientists who conduct basic research and who are interested in fundamental cellular processes are likely to make elemental discoveries about the nature of ES cells as they derive them in the laboratory. Second, significant basic research must be conducted regarding the process of ES cell derivation before cell-based therapies can be fully realized, and this work must be pursued in a wide variety of settings, including those exclusively devoted to basic academic research. Third, ES cells are not indefinitely stable in culture. As these cells are grown, irreversible changes occur in their genetic makeup. Thus, especially in the first few years of human ES cell research, it is important to be able to repeatedly derive ES cells in order to ensure that the properties of the cells that are being studied have not changed.

Thus, anyone who believes that federal support of this important new field of research should maximize its scientific and clinical value within a system of appropriate ethical oversight should be dissatisfied with a position that allows federal agencies to fund research using human ES cells but not research through which the cells are derived from embryos. Instead, recognizing the close connection in practical terms between the derivation and the use of these cells, it would be preferable to enact provisions that apply to funding by all federal agencies, provisions that would carve out a narrow exception for funding of research to use or to derive human ES cells from embryos that would otherwise be discarded by infertility treatment programs.

The Ethical Acceptability of Creating Embryos through IVF Specifcally as a Source of ES Cells

ES cells can be obtained from human research embryos created from donor gametes through IVF for the sole purpose of deriving such cells for research. The primary objection to creating embryos specifically for research is that many believe that there is a morally relevant difference between producing an embryo for the sole purpose of creating a child and producing an embryo with no such goal. Those who object to creating embryos for research often appeal to arguments that speak to respecting human dignity by avoiding the instrumental use of human embryos (i.e., using embryos merely as a means to some other goal does not treat them with appropriate respect or concern as a form of human life). Currently, we believe that cadaveric fetal tissue and embryos remaining after infertility treatments provide an adequate supply of research resources for federal research projects involving human embryos. Therefore, embryos created specifically for research purposes are not needed at the current time in order to conduct important research in this area.

Recommendation 3

Federal agencies should not fund research involving the derivation or use of human ES cells from embryos made solely for research purposes using IVF.

In 1994, the NIH Human Embryo Research Panel argued in support of federal funding of the creation of embryos for research purposes in exceptional cases, such as the need to create banks of cell lines with different genetic makeups that encoded various transplantation antigens—the better to respond, for example, to the transplant needs of groups with different genetic profiles. Such a project would require the recruitment of embryos from genetically diverse donors.

A number of points are worth considering in determining how to deal with this issue. First, it is possible that the creation of research embryos will provide the only opportunity to conduct certain kinds of research—such as research into the process of human fertilization. Second, as IVF techniques improve, it is possible that the supply of embryos for research from this source will dwindle.

Nevertheless, we have concluded that, either from a scientific or a clinical perspective, there is no compelling reason to provide federal funds for the creation of embryos for research at this time.

The Use of SCNT to Obtain ES Cells

The use of SCNT to transfer the nucleus of an adult somatic cell into an enucleated human egg likely has the potential of creating a human embryo. To date, although little is known about these embryos as potential sources of human ES cells, there is significant reason to believe that their use may have therapeutic potential. For example, the possible use of matched tissue for autologous cell-replacement therapy from ES cells may require the use of SCNT. Arguably, the use of this technique to create an embryo is different from the other cases we have considered—because of the asexual origin of the source of the ES cells—although oocyte donation is necessarily involved. We conclude that at this time, because other sources are likely to provide the cells needed for the preliminary stages of research, federal funding should not be provided to derive ES cells from SCNT. Nevertheless, the medical utility and scientific progress of this line of research should be monitored closely.

Recommendation 4

Federal agencies should not fund research involving the derivation or use of human ES cells from embryos made using SCNT into oocytes.

Requirements for the Donation of Cadaveric Fetal Tissue and Embryos for Research

Potential donors of embryos for ES cell research must be able to make voluntary and informed choices about whether and how to dispose of their embryos. Because of concerns about coercion and exploitation of potential donors, as well as controversy regarding the moral status of embryos, it is important, whenever possible, to separate donors' decisions to dispose of their embryos from their decisions to donate them for research. Potential donors should be asked to provide embryos for research only if they have decided to have those embryos discarded instead of donating them to another couple or storing them. If the decision to discard the embryos precedes the decision to donate them for research purposes, then the research determines only how their destruction occurs, not whether it occurs.

Recommendation 5

Prospective donors of embryos remaining after infertility treatments should receive timely, relevant, and appropriate information to make informed and voluntary choices regarding disposition of the embryos. Prior to considering the potential research use of the embryos, a prospective donor should have been presented with the option of storing the embryos, donating them to another woman, or discarding them. If a prospective donor chooses to discard embryos remaining after infertility treatment, the option of donating to research may then be presented. (At any point, the prospective donors' questions—including inquiries about possible research use of any embryos remaining after infertility treatment—should be answered truthfully, with all information that is relevant to the questions presented.)

During the presentation about potential research use of embryos that would otherwise be discarded, the person seeking the donation should
   a) disclose that the ES cell research is not intended to provide medical benefit to embryo donors
   b) make clear that consenting or refusing to donate embryos to research will not affect the quality of any future care provided to prospective donors
   c) describe the general area of the research to be carried out with the embryos and the specific research protocol, if known
   d) disclose the source of funding and expected commercial benefits of the research with the embryos, if known
   e) make clear that embryos used in research will not be transferred to any woman's uterus
   f) make clear that the research will involve the destruction of the embryos

This proposal also stresses the separation that existing laws and policies seek between the pregnant woman's decision to abort and her decision to donate cadaveric fetal tissue for transplantation research. Recommendation 1 proposes to extend that separation to the donation of cadaveric fetal tissue for stem cell research. It may be difficult to achieve this separation in making decisions about embryos that remain after infertility treatments, in part because potential donors at the outset of treatment may have chosen to donate them to research. But, however difficult it may be to achieve, this separation will reduce the chance that potential donors could be pressured or coerced into donating their embryos for stem cell research.

The parts of this recommendation that deal with providing information to donors are designed to ensure that potential donors understand the range of available options and that their decisions are not influenced by anticipated personal medical benefits or by concerns about the quality of subsequent care; that they understand that the research will involve the destruction of the embryos; and that they understand the nature of the proposed research, its source of funding, and its anticipated commercial benefits, if known. Several additional suggested information items are proposed in a document entitled "Points to Consider in Evaluating Basic Research Involving Human Embryonic Stem Cells and Embryonic Germ Cells."

Although the ethical considerations that support the prohibition of the designated donation of human fetal tissue are less acute for EG cell research than they are for transplantation, cause for concern remains. Potential donors of cadaveric fetal tissue for EG cell derivation would not have a direct therapeutic incentive to create or abort tissue for research purposes, as might occur in a transplant context. However, we agree that the prohibition remains a prudent and appropriate way to assure that no incentive—however remote—is introduced into a woman's decision to have an abortion. Any suggestion of personal benefit to the donor or to an individual known to the donor would be untenable and potentially coercive. Thus, the potential donor should be informed both before and after the decision to donate that there is no obligation to make such a gift, that no personal benefit will accrue as a result of the decision to donate, and that no penalty or sanction will result from a decision to refuse to donate.

Recommendation 6

In federally funded research involving embryos remaining after infertility treatments, researchers may not promise donors that ES cells derived from their embryos will be used to treat patient-subjects specified by the donors.

Current provisions regulating fetal tissue research[3] have been narrowly interpreted by NIH and DHHS to apply only where fetal cellular material is intended for transfer into a living human recipient for therapeutic or clinical purposes. No comparable rules exist for human embryos. We believe that this statute should be applied more broadly to include *any* research involving human fetal or embryonic tissue, regardless of its immediate or eventual therapeutic benefit or intended method of intervention. Advances in EG cell research have demonstrated that bioethical concerns are not limited to fetal tissue transplantation.

The Uniform Anatomical Gift Act (UAGA), currently enacted in some form in all fifty states and the District of Columbia, also may require clarification. Current versions of the UAGA explicitly permit donors to make an anatomical gift of the human body or body parts. Because a fetus is included within the UAGA's definition of "decedent," either directly or by implication depending upon the statutory language enacted, the statute's anatomical gift provision undermines any federal prohibition of designated donation of human fetal tissue. What would otherwise qualify for federal statutory preemption is clouded by provisions of the NIH Revitalization Act of 1993 and the federal Common Rule, which direct that fetal tissue transplant researchers must abide by local and state laws, including (by implication) the UAGA.

Finally, if and when sufficient scientific evidence becomes available, clinical benefits are clearly anticipated, and agreement is reached among the various elements in society that the creation of embryos specifically for research or therapeutic purposes is justified (specifically through the use of SCNT), prohibitions on directed donation should be revisited. For obvious reasons, the use of SCNT to develop ES cells for autologous transplantation might require that the recipient be specified.

Prohibitions against the Sale of Embryonic and Fetal Material

Existing rules prohibit the practice of designated donation, the provision of monetary inducements to women undergoing abortion, and the purchase or sale of fetal tissue. We concur in these restrictions and in the recommendation of the 1988 Human Fetal Tissue Transplantation Research Panel that the sale of fetal tissue for research purposes should not be permitted under any circumstances. The potential for coercion is greatest when financial incentives are present, and the treatment of the developing human embryo or fetus as an entity deserving of respect may be greatly undermined by the introduction of any commercial motive into the donation or solicitation of fetal or embryonic tissue for research purposes.

Recommendation 7

Embryos and cadaveric fetal tissue should not be bought or sold.

Policies already in place state that no for-profit trade in fetal tissue should be permitted, and some recommend that the "prohibition on commercial exchange of fetuses and fetal tissue extend to tissues imported from other countries."[4] This prohibition is intended to prevent the exploitation of poor women—especially those in developing countries—who might be persuaded to begin and end pregnancies for money. An important distinction must be made between the possible exploitation of persons that occurs when they are coerced or inappropriately induced to sell parts of their bodies and the exchanges that occur when companies, research institutions, or other groups provide reasonable compensation. This is a familiar issue in discussions about remuneration for participation in research and about which federal regulations defer to Institutional Review Boards (IRBs) for their judgment.

Current regulations[5] attempt to codify this recommendation. Further, depending upon whether a state has enacted the most recent revision of the UAGA (and not all states have enacted the UAGA restriction) and has included the fetus within its definition of "decedent," the sale of fetal remains may or may not be prohibited by individual state statutes. Many states appear to rely on federal statutes and regulations to prohibit fetal tissue sale, and none address human embryos, except by implication.

We strongly encourage those who will draft modified legislation to frame their language in clear terms that are specifically defined. In particular, terms such as "valuable consideration," "processing," and "reasonable payments" require precise definitions.

We believe that, with respect to these regulations, different categories of research intermediaries should be treated differently. One approach would be to establish three intermediary categories: (1) entities responsible for tissue harvest or embryo collection, (2) entities responsible for stem cell derivation or other preresearch preparation *and* postderivation investigators; and (3) *de minimis* intermediaries (including courier or supply services, off-site specimen evaluation, pathological or chemical analysis for research suitability, and other insignificant non- or preresearch patient or specimen interactions). We believe that the first category warrants the greatest degree of regulation. An abortion provider, IVF clinic, or other third party responsible for obtaining consent and/or collecting biological materials should not be able to commercially solicit, pay for, or be paid for the fetal or embryonic material it obtains (permitting only a specifically defined, cost-based reimbursement exception for entities in that category). By placing such prohibitions against paying those who obtain the embryos, it is our intention to discourage the creation of excess embryos during routine infertility procedures, which would later be used for research purposes.

The National Organ Transplant Act (NOTA) prohibition on tissue sale[6] has been criticized for a statutory construction that focuses exclusively on the sale of human organs. Although we agree that fetal organ sale (as well as the sale of embryonic material) should be prohibited, we believe that NOTA's terms are unacceptably narrow and that pre-organ tissues characteristic of early fetal and embryonic development should be included in the tissue sale prohibition.

The Need for National Oversight and Review

The need for national oversight and review of ES and EG cell research is crucial. At present, no such system exists in the United States. A national mechanism to review protocols for deriving human ES and EG cells and to monitor research using such cells would ensure strict adherence to guidelines and standards across the country. Thus, federal oversight can provide the public with the assurance that research involving stem cells is being undertaken appropriately. Given the ethical issues involved in ES and EG cell research—an area in which heightened sensitivity about the very research itself led the president to request that the commission study the issue—the public and the Congress must be assured that oversight can be accomplished efficiently, constructively, in a timely fashion, and with sufficient attention to the relevant ethical considerations.

Several countries, such as the United Kingdom, have recommended the establishment of regulatory boards or national commissions to license and regulate assisted reproductive treatments and embryo research. The use of a national oversight mechanism of this kind has certain advantages, particularly because the use of law to regulate (rather than to set limits) in this area would be burdensome, given the rapid development of biomedical science and technology. On the other hand, some kind of national commission or authority could provide the necessary flexibility and adaptability, and, in addition, such an entity could ensure more consistent ongoing application of safeguards as well as greater public accountability.[7]

In 1994, the NIH Human Embryo Research Panel considered and then explicitly rejected reconstituting the Ethics Advisory Board (EAB) for the purpose of reviewing proposals involving embryos or fertilized eggs:

> Although revisiting the EAB experience offers the potential for public consensus development and a consistent application of the new guidelines, it nonetheless has significant disadvantages. These include the creation of an additional standing government board, the likelihood of significant delay before embryo research could be funded in order to meet legal requirements for new rulemaking prior to the official creation of the government body, and possible further delay if all proposals for embryo research were required to be considered individually by an EAB-type board, despite appearing to be consistent with a developed consensus at NIH about acceptability for funding.[8]

Instead, the NIH panel recommended that

> national review of all protocols by a diverse group of experts is warranted for a time. It is the hope of the Panel that this ad hoc

group will develop additional guidance gained from experience with actual protocols that can be communicated to IRBs through existing mechanisms at NIH.[9]

These recommendations envisioned a time when, following sufficient experience by the ad hoc panel, guidelines for embryo research review could be decentralized to the local IRBs. (It was recommended that the ad hoc panel function for at least three years.) We used similar reasoning in a previous report when recommending that the Secretary of Health and Human Services convene a special standing panel to review certain categories of research involving persons with mental disorders.[10] Like the NIH panel, we did not specify when such guidelines could be decentralized; but unlike the NIH panel, we did recommend that the panel be a standing rather than an ad hoc body.

The NIH panel's recommendations must be viewed in the context of its reporting relationship: The panel was charged with advising the NIH director about research that could be sponsored or conducted by that agency. We note that NIH is not the only federal agency that might be interested in sponsoring or conducting research involving human stem cells; thus, some accommodation must be made for the review of proposals that are not funded by NIH.

Other elements of the NIH panel's recommendation also require additional assessment. For example, the panel recommended that "all such research proposals continue to be specially monitored by the councils and the NIH Office for Protection from Research Risks [OPRR].[11] We are less sanguine than the NIH panel about the ability of OPRR to provide the needed oversight and monitoring for ES and EG cell research at this time, particularly given the recent decision to move this office from NIH to DHHS. Although OPRR's role in the oversight of human subjects research, like that of the FDA, remains central to the structure of human subjects protections in this country, we believe that at this time, an additional mechanism is needed for the review and oversight of federally sponsored research involving human ES and EG cells.

We do, however, share the concern of the 1994 NIH panel, investigators, and IRBs that the process of protocol review should not be viewed as simply a bureaucratic hurdle that researchers must successfully leap solely to satisfy a procedural or regulatory requirement. Done well, protocol review often improves the quality of studies by identifying concerns in the areas of study design, selection of subjects, recruitment, informed consent, and dissemination of results.

Recommendation 8

DHHS should establish a National Stem Cell Oversight and Review Panel to ensure that all federally funded research involving the derivation and/or use of human ES or EG cells is conducted in conformance with the ethical principles and recommendations contained in this report. The panel should have a broad, multidisciplinary membership, including members of the general public, and should

a) review protocols for the derivation of ES and EG cells and approve those that meet the requirements described in this report
b) certify ES and EG cells lines that result from approved protocols
c) maintain a public registry of approved protocols and certified ES and EG cell lines
d) establish a database—linked to the public registry—consisting of information submitted by federal research sponsors (and, on a voluntary basis, by private sponsors, whose proprietary information shall be appropriately protected) that includes all protocols that derive or use ES or EG cells (including any available data on research outcomes, including published papers)
e) use the database and other appropriate sources to track the history and ultimate use of certified cell lines as an aid to policy assessment and formulation
f) establish requirements for and provide guidance to sponsoring agencies on the social and ethical issues that should be considered in the review of research protocols that derive or use ES or EG cells
g) report at least annually to the DHHS secretary with an assessment of the current state of the science for both the derivation and use of human ES and EG cells, a review of recent

developments in the broad category of stem cell research, a summary of any emerging ethical or social concerns associated with this research, and an analysis of the adequacy and continued appropriateness of the recommendations contained in this report.

We recommend several functions that the panel should carry out. In order to accomplish its purposes, the panel should maintain a public registry of federally funded protocols that employ or derive human ES and EG cells and, to the degree possible, a comprehensive listing of privately funded protocols. The purpose of the registry is to make it possible to track not only the protocols themselves and their adherence to the principles described above, but also their outcomes and the outcomes of all research based on their results. The panel should be able to describe the history and trajectory of research that uses these cells and to guard against the promiscuous use of the cells. As they are submitted, new federally funded protocols involving the derivation of ES cells must include a statement that only certified cell lines will be used.

Knowledge about the history and ultimate outcome and use of research using human ES and EG cells should be open to the public. Thus, the information accumulated by the panel through the registry should be used not only for ethical review, but also for public education. This is an important educational and informational function that may encourage the active participation of the private sector in the registry—even in the absence of any federal regulatory requirement to do so. In addition, within five years, the panel and the registry should be independently reviewed. This review, which should include an evaluation of the processes of the oversight and review mechanisms, will help to determine whether the level of limitations on this area of research is appropriate as well as to determine whether case-by-case review of derivation protocols is still warranted.

There are several benefits to a national review process for all federally funded research on ES and EG cells. These include preventing ethically problematic research, assuring the public that the research is scientifically meritorious and ethically acceptable; providing information by which to evaluate issues of social justice in the use of the knowledge or other products of the research; providing public oversight of controversial research practices; assuring consistency in the review of protocols; evaluating this type of research; and educating the public.

Although we are aware that NIH will likely conduct and/or fund the majority of federally sponsored stem cell research in the country and will be developing its own set of guidelines for the conduct of ES and EG cell research, we are persuaded that it is important to distance to some degree the review and oversight of stem cell research from what is the principal source of funding in this country. The proximity of NIH within DHHS (our recommended location for the panel) makes it possible for a number of beneficial arrangements to develop. These include developing requirements for data sharing as a condition for receiving grants; developing guidelines for sharing cell lines; and providing a common review mechanism for other federal agencies that are conducting/funding research involving ES and EG cells (e.g., through a memorandum of understanding).

The Need for Local Review of Derivation Protocols

For more than two decades, prospective review by an IRB has been the principal method for assuring that federally sponsored research involving human subjects will be conducted in compliance with guidelines, policies, and regulations designed to protect human beings from harm. This system of local review has been subject to criticism, and, indeed, in previous analyses we have identified a number of concerns regarding this system of review. In preparing this report, we considered a number of proposals that would allow for the local review of research protocols involving human ES and EG cell research, bearing in mind that a decision by the commission to recommend a role for IRBs might be incorrectly interpreted as endorsing the view that human ES or EG cells or human embryos are human subjects.

We adopted the principle, reflected in these recommendations, that for research involving the derivation of ES and EG cells, a system of national oversight and review would be needed to ensure that important federal sponsorship of stem cell research could proceed—but only under specific conditions. We recognized that for such research proposals, many of the ethical issues could be considered at the local level—that is, at the institutions where the research would be conducted. In general, the IRB is an appropriate body for reviewing protocols that aim to derive ES or EG cells. Although few review bodies (including IRBs) have extensive experience in the review of such protocols, IRBs remain the most visible and expert entities available. It is for this reason, for example, that a number of recommendations presented in this report (8, 9, 10, 11, and 12) discuss the importance of developing additional guidance for the review of protocols that involve human stem cell research.

Recommendation 9

Protocols involving the derivation of human ES and EG cells should be reviewed and approved by an IRB or by another appropriately constituted and convened institutional review body prior to consideration by the National Stem Cell Oversight and Review Panel. (See Recommendation 8.) This review should ensure compliance with any requirements established by the panel, including confirming that individuals or organizations (in the United States or abroad) that supply embryos or cadaveric fetal tissue have obtained them in accordance with the requirements established by the panel.

As noted earlier, for research proposals that involve the derivation of human ES or EG cells, particular ethical issues require attention through a national review process. However, this process should begin at the local level, because institutions that intend to conduct research involving the derivation of human ES cells or EG cells should continue to take responsibility for ensuring the ethical conduct of that research. More important, however, IRBs can play an important role—particularly by reviewing consent documents and by assuring that collaborative research undertaken by investigators at foreign institutions has satisfied any regulatory requirements for the sharing of research materials.

Currently there is no definitive answer to the question of whether the Common Rule, 45 CFR 46, and/or Subpart B apply to research involving fetal tissue transplantation, to human embryo research, and by extension to EG and ES cell research. If the regulations do apply, then IRBs would be expected to review protocols, consistent with the regulatory requirements. We have indicated, however, that even if these regulations do not apply, we believe that IRBs should be expected to review derivation protocols to assess their ethical acceptability without having to commit to a position that the activities are human subjects research as defined by the regulations. If, as a matter of public policy, ES and EG cell research were found to be human subjects research, certain clarifying changes in the regulations might be needed. For example, Subpart B would need to provide clearly that *any* living donor of human biological material constitutes a human subject for purposes of research protection, and IRB review and informed consent under all subparts of the DHHS version of the Common Rule would need to apply. Similarly, we have made clear that the authorization a woman may give to donate fetal tissue following an elective abortion may better be understood as consent to donate—analogous to donating organs—rather than as providing informed consent for research participation. Even if these models differ, the principle we adopt remains the same: opportunity for consent should rest exclusively with the individual or individuals legally empowered to assume a donative role.[12]

Responsibilities of Federal Research Agencies

We have recommended that protocols involving the *derivation* of ES or EG cells should be reviewed by both a local review group and the national panel described in Recommendation 8. For protocols that involve only the use but not the derivation of ES or EG cells, oversight and review are still necessary, but these protocols do not require reliance on such a system. In our judgment, these protocols can be appropriately reviewed using the existing system for the submission, review, and approval of research proposals that is in place at federal research agencies, which includes the use of a peer review group—sometimes called a study section or initial review group—that is established to assess the scientific merit of the proposals. In addition, in some agencies, such as NIH, staff members review protocols before they are transmitted to a national advisory council for final approval. These levels of review all provide an opportunity to consider ethical issues that arise in the proposals. When research proposals involve human subjects, in order to assure that it is ethically acceptable, federal agencies rely on local IRBs for review and approval. (See Recommendation 9.) At every point in this continuum—from the first discussions that a prospective applicant may have with program staff within a particular institute to the final decision by the relevant national advisory council—ethical and scientific issues can be addressed by the sponsoring agency. But even if—based on a particular interpretation of the federal regulation—these research proposals do not involve human subjects, we believe the system of oversight and review can adequately address the relevant ethical issues.

Recommendation 10

All federal agencies should ensure that their review processes for protocols using human ES or EG cells comply with any requirements established by the National Stem Cell Oversight and Review Panel (see Recommendation 8), paying particular attention to the adequacy of the justification for using such cell lines.

Research involving human ES and EG cells raises critical ethical issues, particularly when the proposals involve the derivation of ES cells from embryos that remain after infertility treatments. We recognize that these research proposals may not follow the paradigm that is usually associated with human subjects research. Nevertheless, research proposals that are being considered for funding by federal agencies must, in our view, meet the highest standards of scientific merit and ethical acceptability. To that end, the recommendations made in this report, including a proposed set of points to consider in evaluating basic research involving human ES cells and EG cells, constitute a set of ethical and policy considerations that should be reflected in the respective policies of federal agencies conducting or sponsoring human ES or EG cell research.

Attention to Issues for the Private Sector

Although this report primarily addresses the ethical issues associated with the use of federal funds for research involving the derivation and/or use of ES and EG cells, we recognize that considerable work in both of these areas will be conducted under private sponsorship. Thus, our recommendations may have implications for those working in the private sector. First, for cell lines to be eligible for use in federally funded research, they must be certified by the National Stem Cell Oversight and Review Panel described in Recommendation 8. Therefore, if a private company aims to make its cell lines available to publicly funded researchers, it must submit its derivation protocol(s) to the same oversight and review process recommended for the public sector, (i.e., local review; see Recommendation 9) and for certification by the proposed national panel that the cells have been derived from embryos remaining after infertility treatments or from cadaveric fetal tissue.

Second, we hope that nonproprietary aspects of protocols developed under private sponsorship will be made available in the public registry, as described in Recommendation 8. The greater the participation of the private sector in providing information on human ES and EG cell research, the more comprehensive the development of the science and related public policies in this area.

Third, and perhaps most relevant in an ethically sensitive area of emerging biomedical research, it is important that all members of the research community, whether in the public or private sector, conduct the research in a manner that is open to appropriate public scrutiny. During the last two decades, we have witnessed an unprecedented level of cooperation between the public and private sectors in biomedical research, which has resulted in the international leadership position of the United States in this area. Public bodies and other authorities, such as the Recombinant DNA Advisory Committee, have played a crucial role in enabling important medical advances in fields such as gene therapy by providing oversight of both publicly and privately funded research efforts. We believe that voluntary participation by the private sector in the review and certification procedures of the proposed national panel, as well as in its deliberations, can contribute equally to the socially responsible development of ES and EG cell technologies and accelerate their translation into biomedically important therapies that will benefit patients.

Recommendation 11

For privately funded research projects that involve ES or EG cells that would be eligible for federal funding, private sponsors and researchers are encouraged to adopt voluntarily the applicable recommendations of this report. This includes submitting protocols for the derivation of ES or EG cells to the National Stem Cell Oversight and Review Panel for review and cell line certification. (See Recommendations 8 and 9.)

In this report, we recommend that federally funded research that involves the derivation of ES cells should be limited to those efforts that use embryos that remain after infertility treatments. Some of the recommendations made in this context—such as the requirement for separating the decision by a woman to cease infertility treatment when embryos still remain from her decision to donate those embryos to research—simply do not apply to efforts to derive ES cells from embryos created (whether by IVF or by SCNT) solely for research purposes—activities that might be pursued in the private sector. Nevertheless, other ethical standards and safeguards embodied in the recommendations, such as provisions to prevent the coercion of women and the promotion of commerce in human reproduction, remain vitally important, even when embryos are created solely for research purposes.

Recommendation 12

For privately funded research projects that involve deriving ES cells from embryos created solely for research purposes and that are therefore not eligible for federal funding (see Recommendations 3 and 4).

   a) professional societies and trade associations should develop and promulgate ethical safeguards and standards consistent with the principles underlying this report
   b) private sponsors and researchers involved in such research should voluntarily comply with these safeguards and standards

Professional societies and trade associations dedicated to reproductive medicine and technology play a central role in establishing policy and standards for clinical care, research, and education. We believe that these organizations can and should play a salutary role in ensuring that all embryo research conducted in the United States, including that which is privately funded, conforms to the ethical principles underlying this report. Many of these organizations already have developed policy statements, ethics guidelines, or other directives addressing issues in this report, and we have bene-fited from a careful review of these materials. These organizations are encouraged to review their professional standards to ensure not only that they keep pace with the evolving science of human ES and EG cell research, but also that their members are knowledgeable about and in compliance with them. For those organizations that conduct research in this area but that lack statements or guidelines addressing the topics of this report, we recommend strongly that they develop such statements or guidelines. No single institution or organization, whether in the public or the private sector, can pro-vide all the necessary protections and safeguards.

The Need for Ongoing Review and Assessment

No system of federal oversight and review of such a sensitive and important area of investigation should be established without at the same time providing an evaluation of its effectiveness, value, and ongoing need. The pace of scientific development in human ES and EG cell research likely will increase. Although one cannot predict the direction of the science of human stem cell research, in order for the American public to realize the promise of this research and to be assured that it is being conducted responsibly, close attention to and monitoring of all the mechanisms established for oversight and review are required.

Recommendation 13

The National Stem Cell Oversight and Review Panel described in Recom-mendation 8 should be chartered for a fixed period of time, not to exceed five years. Prior to the expiration of this period, DHHS should commission an independent evaluation of the panel's activities to determine whether it has adequately fulfilled its functions and whether it should be continued.

There are several reasons for allowing the national panel to function for a fixed period of time and for evaluating its activities before it continues its work. First, some of the hoped-for results will be avail-able from research projects that are using the two sources we con-sider to be ethically acceptable for federal funding. Five years is a reasonable period in which to allow some of this information to amass, offering the panel, researchers, members of Congress, and the public sufficient time to determine whether any of the knowledge or potential health benefits are being realized. The growing body of information in the public registry and database described above (particularly if privately funded researchers and sponsors voluntarily participate) will aid these considerations.

Second, within this period the panel may be able to determine whether additional sources of ES cells are necessary in order for important research to continue. Two arguments have been offered for supporting research using embryos created specifically for research purposes: One is the concern that not enough embryos remain for this purpose from infertility treatments, and the other is the recognition that some research requires embryos that are generated for specific research and/or medical purposes. The panel should assess whether additional sources of ES cells that we have judged to be ineligible for federal funding at this time (i.e., embryos created solely for research purposes) are legitimately needed.

Third, an opportunity to assess the relationship between local review of protocols using human ES and EG cells and the panel's review of protocols for the derivation of ES cells will be offered. It will, of course, take time for this national oversight and review mechanism to develop experience with the processes of review, certification, and approval described in this report.

Fourth, we hope that the panel will contribute to the broad and ongoing national dialogue on the ethical issues regarding research involving human embryos. A recurring theme of our deliberations, and in the testimony we heard, was the importance of encouraging this national conversation.

The criteria for determining whether the panel has adequately fulfilled its functions should be set forth by an independent body established by DHHS. However, it would be reasonable to expect that the evaluation would rely generally on the seven functions described above in Recommendation 8 and that this evaluation would be conducted by a group with the requisite expertise. In addition, some of the following questions might be considered when conducting this evaluation: Is there reason to believe that the private sector is voluntarily submitting descriptions of protocols involving the derivation of human ES cells to the panel for review? Is the panel reviewing projects in a timely manner? Do researchers find that the review process is substantively helpful? Is the public being provided with the assurance that social and ethical issues are being considered?

SUMMARY

Recent developments in human ES and EG cell research have raised the prospect that new therapies will become available that will serve to relieve human suffering. These developments also have served to remind society of the deep moral concerns that are related to research involving human embryos and cadaveric fetal tissue. Serious ethical discussion will (and should) continue on these issues. However, in light of public testimony, expert advice, and published writings, we have found substantial agreement among individuals with diverse perspectives that although the human embryo and fetus deserve respect as forms of human life, the scientific and clinical benefits of stem cell research should not be foregone. We were persuaded that carrying out human ES cell research under federal sponsorship is important, but only if it is conducted in an ethically responsible manner. After extensive deliberation, the commission believes that acceptable public policy can be forged, in part, based upon these widely shared views. Through this report, we not only offer recommendations regarding federal funding and oversight of stem cell research, but also hope to further stimulate the important public debate about the profound ethical issues regarding this potentially beneficial research.

NOTES

1. Use of fetal tissue in research is also permitted in Canada, the United Kingdom, Australia, and in most countries in the European Union. Germany, for example, does not permit embryo research but does permit the use of fetal tissue for the derivation of EG cells. The German statement concerning human ES cells upholds the ban on destructive embryo research, effectively banning the derivation of ES cells, because the option of deriving EG cells exists in that country. See the German statement concerning the question of human ES cells, Deutsche Forschungsgemeinschaf,

"Statement Concerning the Question of Human Embryonic Stem Cells" [online], www.dfg-bonn.de/english/ press/eszell_e.html [March 1999], pp. 8–10.

2. Public Law 105-78, 513(a) (1997).

3. 42 USC § 289g-1 and g-2

4. Canadian Royal Commission on New Reproductive Technologies, *Proceed with Care: Final Report of the Royal Commission on New Reproductive Technologies*, 2 vols. (Ottawa: Minister of Government Services, 1993).

5. 42 USC §§ 289g-2(a), 289g-2(b)(3), 274e, and 42 CFR § 46.206(b).

6. 42 USC § 274e.

7. European Group on Ethics, "Opinion of the European Group on Ethics in Science and New Technologies,"1998, at art. 2.11; see also the Australian NHMRC *Guidelines* (1996) advocating that complementary nationally assisted reproductive technology standards or legislation be adopted in the Australian states.

8. National Institutes of Health (NIH) Human Embryo Research Panel, *Report of the Human Embryo Research Panel*, 2 vols. (Bethesda, Md.: NIH, 1994), p. 72.

9. Ibid., p. 73.

10. National Bioethics Advisory Commission, *Research Involving Persons with Mental Disorders That May Affect Decisionmaking Capacity*, 2 vols. (Rockville, Md.: U.S. Government Printing Office, 1998).

11. NIH Human Embryo Research Panel, p. 74.

12. See *Federal Register* 27804, proposed rule 45 CFR § 46.204(d)-(e) and Table 1, "Current and Proposed 45 CFR 46, Subpart B," 27798, explanatory text ("consent of the father is not required"; rather, "consent of the mother or her legally authorized representative is required" [after she is]…"informed of the reasonably foreseeable impact of the research on the fetus").

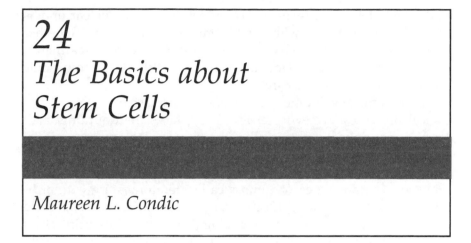

24
The Basics about Stem Cells

Maureen L. Condic

I n August of last year, President Bush approved the use of federal funds to support research on a limited number of existing human embryonic stem cell lines. The decision met with notably mixed reactions. Proponents of embryonic stem cell research argue that restricting federal funding to a limited number of cell lines will hamper the progress of science, while those opposed insist that any use of cells derived from human embryos constitutes a significant breach of moral principles. It is clear that pressure to expand the limits established by the president will continue. It is equally clear that the ethical positions of those opposed to this research are unlikely to change.

Regrettably, much of the debate on this issue has taken place on emotional grounds, pitting the hope of curing heartrending medical conditions against the deeply held moral convictions of many Americans. Such arguments frequently ignore or mischaracterize the scientific facts. To arrive at an informed opinion on human

Reprinted from Maureen L. Condic, "The Basics about Stem Cells," *First Things* (January 2002): 30–34.

embryonic stem cell research, it is important to have a clear understanding of precisely what embryonic stem cells are, whether embryonic stem cells are likely to be useful for medical treatments, and whether there are viable alternatives to the use of embryonic stem cells in scientific research.

Embryonic development is one of the most fascinating of all biological processes. A newly fertilized egg faces the daunting challenge of not only generating all of the tissues of the mature animal but organizing them into a functionally integrated whole. Generating a wide range of adult cell types is not an ability unique to embryos. Certain types of tumors called teratomas are extraordinarily adept at generating adult tissues, but unlike embryos, they do so without the benefit of an organizing principle or blueprint. Such tumors rapidly produce skin, bone, muscle, and even hair and teeth, all massed together in a chaotic lump of tissue. Many of the signals required to induce formation of specialized adult cells must be present in these tumors, but unlike embryos, tumors generate adult cell types in a hopelessly undirected manner.

If a developing embryo is not to end up a mass of disorganized tissues, it must do more than generate adult cell types. Embryos must orchestrate and choreograph an elaborate stage production that gives rise to a functional organism. They must direct intricate cell movements that bring together populations of cells only to separate them again, mold and shape organs through the birth of some cells and the death of others, and build ever more elaborate interacting systems while destroying others that serve only transient, embryonic functions. Throughout the ceaseless building, moving, and remodeling of embryonic development, new cells with unique characteristics are constantly being generated and integrated into the overall structure of the developing embryo. Science has only the most rudimentary understanding of the nature of the blueprint that orders embryonic development. Yet, recent research has begun to illuminate both how specific adult cells are made as well as the central role of stem cells in this process.

The term "stem cell" is a general one for any cell that has the ability to divide, generating two progeny (or "daughter cells"), one of which is destined to become something new and one of which replaces the original stem cell. In this sense, the term "stem" identifies these cells as the source or origin of other, more specialized cells.

There are many stem cell populations in the body at different stages of development. For example, all of the cells of the brain arise from a neural stem cell population in which each cell produces one brain cell and another copy of itself every time it divides. The very earliest stem cells, the immediate descendants of the fertilized egg, are termed embryonic stem cells, to distinguish them from populations that arise later and can be found in specific tissues (such as neural stem cells). These early embryonic stem cells give rise to all the tissues in the body, and are therefore considered "totipotent" or capable of generating all things.

While the existence of early embryonic stem cells has been appreciated for some time, the potential medical applications of these cells have only recently become apparent. More than a dozen years ago, scientists discovered that if the normal connections between the early cellular progeny of the fertilized egg were disrupted, the cells would fall apart into a single cell suspension that could be maintained in culture. These dissociated cells (or embryonic stem cell "lines") continue to divide indefinitely in culture. A single stem cell line can produce enormous numbers of cells very rapidly. For example, one small flask of cells that is maximally expanded will generate a quantity of stem cells roughly equivalent in weight to the entire human population of the earth in less than sixty days. Yet despite their rapid proliferation, embryonic stem cells in culture lose the coordinated activity that distinguishes embryonic development from the growth of a teratoma. In fact, these early embryonic cells in culture initially appeared to be quite unremarkable: a pool of identical, relatively uninteresting cells.

First impressions, however, can be deceiving. It was rapidly discovered that dissociated early embryonic cells retain the ability to generate an astounding number of mature cell types in culture if they are provided with appropriate molecular signals. Discovering the signals that induce the formation of specific cell types has been an arduous task that is still ongoing. Determining the precise nature of the cells generated from embryonic stem cells has turned out to be a matter of considerable debate. It is not at all clear, for example, whether a cell that expresses some of the characteristics of a normal brain cell in culture is indeed "normal"—that is, if it is fully functional and capable of integrating into the architecture of the brain without exhibiting any undesirable properties (such as malignant growth). Nonetheless, tremendous excitement accompanied the discovery of dissociated cells' generative power, because it was widely believed that cultured embryonic stem cells would retain their totipotency and could therefore be induced to generate all of the mature cell types in the body. The totipotency of cultured embryonic stem cells has not been demonstrated and would, in fact, be difficult to prove. Nonetheless, because it is reasonable to assume embryonic stem cells in culture retain the totipotency they exhibit in embryos, this belief is held by many as an article of faith until proven otherwise.

Much of the debate surrounding embryonic stem cells has centered on the ethical and moral questions raised by the use of human embryos in medical research. In contrast to the widely divergent public opinions regarding this research, it is largely assumed that from the perspective of science there is little or no debate on the matter. The scientific merit of stem cell research is most commonly characterized as "indisputable" and the support of the scientific community as "unanimous." Nothing could be further from the truth. While the scientific advantages and potential medical application of embryonic stem cells have received considerable attention in the public media, the equally compelling scientific and medical disadvantages of transplanting embryonic stem cells or their derivatives into patients have been ignored.

There are at least three compelling scientific arguments against the use of embryonic stem cells as a treatment for disease and injury. First and foremost, there are profound immunological issues associated with putting cells derived from one human being into the body of another. The same compromises and complications associated with organ transplant hold true for embryonic stem cells. The rejection of transplanted cells and tissues can be slowed to some extent by a good "match" of the donor to the patient, but except in cases of identical twins (a perfect match), transplanted cells will eventually be targeted by the immune system for destruction. Stem cell transplants, like organ transplants, would not buy you a "cure"; they would merely buy you time. In most cases, this time can only be purchased at the dire price of permanently suppressing the immune system.

The proposed solutions to the problem of immune rejection are either scientifically dubious, socially unacceptable, or both. Scientists have proposed large scale genetic engineering of embryonic stem cells to alter their immune characteristics and provide a better match for the patient. Such a manipulation would not be trivial; there is no current evidence that it can be accomplished at all, much less as a safe and routine procedure for every patient. The risk that genetic mutations would be introduced into embryonic stem cells by genetic engineering is quite real, and such mutations would be difficult to detect prior to transplant.

Alternatively, the use of "therapeutic cloning" has been proposed. In this scenario, the genetic information of the original stem cell would be replaced with that of the patient, producing an embryonic copy or "clone" of the patient. This human clone would then be grown as a source of stem cells for transplant. The best scientific information to date from animal cloning experiments indicates that such "therapeutic" clones are highly likely to be abnormal and would not give rise to healthy replacement tissue.

The final proposed resolution has been to generate a large bank of embryos for use in transplants. This would almost certainly involve the creation of human embryos with specific immune characteristics ("Wanted: sperm donor with AB+ blood type") to fill in the "holes" in our collection. Intentionally producing large numbers of human embryos solely for scientific and medical use is not an option most people would be willing to accept. The three proposed solutions to the immune problem are thus no solution at all.

The second scientific argument against the use of embryonic stem cells is based on what we know about embryology. In an opinion piece published in the *New York Times*[1] a noted stem cell researcher, Dr. David Anderson, relates how a seemingly insignificant change in "a boring compound" that allows cells to stick to the petri dish proved to be critical for inducing stem cells to differentiate as neurons. There is good scientific reason to believe the experience Dr. Anderson describes is likely to be the norm rather than a frustrating exception. Many of the factors required for the correct differentiation of embryonic cells are not chemicals that can be readily "thrown into the bubbling cauldron of our petri dishes." Instead, they are structural or mechanical elements uniquely associated with the complex environment of the embryo.

Cells frequently require factors such as mechanical tension, large scale electric fields, or complex structural environments provided by their embryonic neighbors in order to activate appropriate genes and maintain normal gene-expression patterns. Fully reproducing these nonmolecular components of the embryonic environment in a petri dish is not within the current capability of experimental science, nor is it likely to be so in the near future. It is quite possible that even with "patience, dedication, and financing to support the work," we will never be able to replicate in a culture dish the nonmolecular factors required to get embryonic stem cells "to do what we want them to."

Failing to replicate the full range of normal developmental signals is likely to have disastrous consequences. Providing some but not all of the factors required for embryonic stem cell differentiation could readily generate cells that appear to be normal (based on the limited knowledge scientists have of what constitutes a "normal cell type") but are in fact quite abnormal. Transplanting incompletely differentiated cells runs the serious risk of introducing cells with abnormal properties into patients. This is of particular concern in light of the enormous tumor-forming potential of embryonic stem cells. If only one out of a million transplanted cells somehow failed to receive the correct signals for differentiation, patients could be given a small number of fully undifferentiated embryonic stem cells as part of a therapeutic treatment. Even in very small numbers, embryonic stem cells produce teratomas, rapidly growing and frequently lethal tumors. (Indeed, formation of such tumors in animals is one of the scientific assays for the "nultipotency" of embryonic stem cells.) No currently available level of quality control would be sufficient to guarantee that we could prevent this very real and horrific possibility.

The final argument against using human embryonic stem cells for research is based on sound scientific practice: We simply do not have sufficient evidence from animal studies to warrant a move to human experimentation. While there is considerable debate over the moral and legal status of early human embryos, this debate in no way constitutes a justification to step outside the normative practice of science and medicine that requires convincing and reproducible evidence from animal models prior to initiating experiments on (or, in this case, with) human beings. While the "potential promise" of embryonic stem cell research has been widely touted, the data supporting that promise is largely nonexistent.

To date there is *no* evidence that cells generated from embryonic stem cells can be safely transplanted back into adult animals to restore the function of damaged or diseased adult tissues. The level of scientific rigor that is normally applied (indeed, legally required) in the development of potential medical treatments would have to be entirely ignored for experiments with human embryos to proceed. As our largely disappointing experience with gene therapy should remind us, many highly vaunted scientific techniques frequently fail to yield the promised results. Arbitrarily waiving the requirement for scientific evidence out of a naive faith in "promise" is neither good science nor a good use of public funds.

Despite the serious limitations to the potential usefulness of embryonic stem cells, the argument in favor of this research would be considerably stronger if there were no viable alternatives. This, however, is decidedly not the case. In the last few years, tremendous progress has been made in the field of adult stem cell research. Adult stem cells can be recovered by tissue biopsy from patients, grown in culture, and induced to differentiate into a wide range of mature cell types.

The scientific, ethical, and political advantages of using adult stem cells instead of embryonic ones are significant. Deriving cells from an adult patient's own tissues entirely circumvents the problem of immune rejection. Adult stem cells do not form teratomas. Therapeutic use of adult stem cells raises very few ethical issues and completely obviates the highly polarized and acrimonious political debate associated with the use of human embryos. The concern that cells derived from diseased patients may themselves be abnormal is largely unwarranted. Most human illnesses are caused by injury or by foreign agents (toxins, bacteria, viruses, etc.) that, if left untreated, would affect adult and embryonic stem cells equally. Even in the minority of cases where human illness is caused by genetic factors, the vast majority of such illnesses occur relatively late in the patient's life. The late onset of genetic diseases suggests such disorders would take years or even decades to re-emerge in newly generated replacement cells.

In light of the compelling advantages of adult stem cells, what is the argument against their use? The first concern is a practical one: Adult stem cells are more difficult than embryonic ones to grow in culture and may not be able to produce the very large numbers of cells required to treat large numbers of patients. This is a relatively trivial objection for at least two reasons. First, improving the proliferation rate of cells in culture is a technical problem that science is quite likely to solve in the future. Indeed, substantial progress has already been made toward increasing the rate of adult stem cell proliferation. Second, treating an individual patient using cells derived from his own tissue ("autologous transplant") would not require the large numbers of cells needed to treat large populations of patients. A slower rate of cell proliferation is unlikely to prevent adult stem cells from generating sufficient replacement tissue for the treatment of a single patient.

The more serious concern is that scientists don't yet know how many mature cell types can be generated from a single adult stem cell population. Dr. Anderson notes, "Some experiments suggest these [adult] stem cells have the potential to make midcareer switches, given the right environment, but in most cases this is far from conclusive." This bothersome limitation is not unique to adult stem cells. Dr. Anderson goes on to illustrate that in most cases the evidence suggesting scientists can induce embryonic stem cells to follow a specific career path is equally far from conclusive. In theory, embryonic stem cells appear to be a more attractive option because they are clearly capable (in an embryonic environment) of generating all the tissues of the human body. In practice, however, it is extraordinarily difficult to get stem cells of any age "to do what you want them to" in culture.

There are two important counterarguments to the assertion that the therapeutic potential of adult stem cells is less than that of embryonic stem cells because adult cells are "restricted" and therefore unable to generate the full range of mature cell types. First, it is not clear at this point whether adult stem cells are more restricted than their embryonic counterparts. It is important to bear in mind that the field of adult stem cell research is not nearly as advanced as the field of embryonic stem cell research. Scientists have been working on embryonic stem cells for more than a decade, whereas adult stem cells have only been described within the last few years. With few exceptions, adult stem cell research has demonstrated equal or greater promise than embryonic stem cell research at a comparable stage of investigation. Further research may very well prove that it is just as easy to teach an old dog new tricks as it is to train a willful puppy. This would not eliminate the very real problems associated with teaching any dog to do anything useful, but it would remove the justification for "age discrimination" in the realm of stem cells.

The second counterargument is even more fundamental. Even if adult stem cells are unable to generate the full spectrum of cell types found in the body, this very fact may turn out to be a strong scientific and medical advantage. The process of embryonic development is a continuous trade-off between potential and specialization. Embryonic stem cells have the potential to become anything, but are specialized at nothing. For an embryonic cell to specialize, it must make choices that progressively restrict what it can become. The greater the number of steps required to achieve specialization, the greater the scientific challenge it is to reproduce those steps in culture. Our current understanding of embryology is nowhere near advanced enough for scientists to know with confidence that we have gotten all the steps down correctly. If adult stem cells prove to have restricted rather than unlimited potential, this would indicate that adult stem cells have proceeded at least part way toward their final state, thereby reducing the number of steps scientists are required to replicate in culture. The fact that adult stem cell development has been directed by nature rather than by scientists greatly increases our confidence in the normalcy of the cells being generated.

There may well be multiple adult stem cell populations, each capable of forming a different subset of adult tissues, but no one population capable of forming everything. This limitation would make certain scientific enterprises considerably less convenient. However, such a restriction in "developmental potential" would not limit the therapeutic potential of adult stem cells for treatment of disease and injury. Patients rarely go to the doctor needing a full body replacement. If a patient with heart disease can be cured using adult cardiac stem cells, the fact that these "heart restricted" stem cells do not generate kidneys is not a problem for the patient.

The field of stem cell research holds out considerable promise for the treatment of disease and injury, but this promise is not unlimited. There are real, possibly insurmountable, scientific challenges to the use of embryonic stem cells as a medical treatment for disease and injury. In contrast, adult stem cell research holds out nearly equal promise while circumventing the enormous social, ethical, and political issues raised by the use of human embryos for research. There is clearly much work that needs to be done before stem cells of any age can be used as a medical treatment. It seems only practical to put our resources into the approach that is most likely to be successful in the long run. In light of the serious problems associated with embryonic stem cells and the relatively unfettered promise of adult stem cells, there is no compelling scientific argument for the public support of research on human embryos.

Jerusalem Artichoke

It's *not*, though: Anyone can tell you this is
absolutely not an artichoke.
And the *Jerusalem* that it professes?
Some Italian costermonger spoke
about the sunflower—the English heard
not *girasole* but, *Jerusalem*—
and, naming these vile tubers, they conferred
an accidental dignity on them.

People say a flower shows us
certain proof that there's a God;
is *Helianthus tuberosus'*
ugly tuber—sunk in sod—
proof of the devil? If the flower
needs the root, the devil's sent
expressly by a higher power:
Neither thing's an accident.

Deborah Warren

NOTE

1. David Anderson, "The Alchemy of Stem Cell Research," *New York Times*, July 15, 2001, sec. 4, p. 15, col. 1 (Editorial).

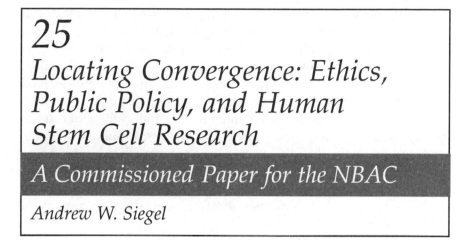

25
Locating Convergence: Ethics, Public Policy, and Human Stem Cell Research

A Commissioned Paper for the NBAC

Andrew W. Siegel

INTRODUCTION: BENEFITS AND CONSTRAINTS

The principal moral justification for promoting research with human pluripotent stem cells is that such research has the potential to lead to direct health benefits to individuals suffering from disease. Research that identifies the mechanisms controlling cell differentiation would provide the foundation for directed differentiation of pluripotent stem cells to specific cell types. The ability to direct the differentiation of stem cells would, in turn, advance the development of therapies for repairing injuries and pathological processes. The great promise of human stem cell research inspired thirty-three Nobel laureates to voice their support for the research and to lay down the gauntlet against those who oppose it: "Those who seek to prevent medical advances using stem cells must be held accountable to those who suffer from horrible disease and their families, why such hope should be withheld."[1]

While the invocation of the potential benefits of stem cell research furnishes strong moral grounds for supporting the research, considerations of social utility are not always sufficient to morally justify actions. There are moral constraints on the promotion of the social good. For example, considerations of justice and respect for persons often trump considerations of social utility. Those who oppose research involving the use of stem cells derived from embryos and fetal tissue argue that the research is morally impermissible because it is implicated in the unjust killing of human beings who have the moral status of persons. Opponents of the research maintain that the constraints against killing persons to advance the common good apply to fetuses and embryos. On this view, one cannot approve of a policy supporting such research without holding that elective abortion and the destruction of embryos are morally permissible.

In this paper, I consider whether it is possible to offer a justification for federal support of research with stem cells derived from fetal tissue and embryos that is not premised on a view about the morality of abortion and the destruction of embryos. The aim of seeking such a justification is to provide a foundation for consensus on public policy in the area of stem cell research. I will argue that people with divergent views about the moral status of fetuses and embryos should be able to endorse some research uses of stem cells derived from these sources.

RESEARCH WITH STEM CELLS DERIVED FROM FETAL TISSUE

The ethical acceptability of deriving stem cells from the tissue of aborted fetuses is, for some, closely connected to the morality of elective abortion. For those who believe that abortion is permissible because the fetus has no moral standing, there are no significant moral barriers to research using stem cells derived from fetal tissue. Restrictions that separate decisions to donate fetal tissue from decisions to abort might be thought necessary, but the purpose of the restrictions would be to protect the mother against coercion and exploitation rather than to protect the fetus.

What is less clear is whether one can both morally oppose abortion and support this method of deriving stem cells. A common view in the literature on the ethics of human fetal tissue transplantation research is that we can support the research without assuming that abortion is morally permissible. As long as guidelines are in place that ensure that abortion decisions and procedures are separated from considerations of fetal tissue procurement and use in research, using aborted fetuses for research is no more problematic than using other cadavers donated for scientific and medical purposes.

Opponents of the research use of fetal materials obtained from induced abortions dispute the claim that we can dismiss the relevance of the morality of abortion. They appeal to two grounds in asserting the relevance of the moral question: (1) those who procure and use fetal material from induced abortions are complicit with the abortions which provide the material, and (2) it is impossible to obtain valid informed consent for the use of fetal materials. Each of these claims merits consideration.

Complicity

There are two general ways in which one can maintain that those involved in the research use of stem cells derived from aborted fetuses are complicit with abortion: (1) They bear some causal responsibility for abortions, or (2) they symbolically align themselves with abortion.

Causal Responsibility

A researcher or tissue procurer may bear direct or indirect causal responsibility for abortions. The former kind of responsibility exists where one directly motivates a woman to have an abortion (e.g., by offering financial incentives) or is directly involved in the performance of an abortion from which fetal tissue is procured. There are several measures that can help prevent direct responsibility for abortions:

- A requirement that the consent of women for abortions be obtained prior to requesting or obtaining consent for the donation of fetal tissue
- A requirement that those who seek a woman's consent to donate not discuss fetal tissue donation prior to her decision to donate
- A prohibition against the sale of fetal tissue

- A prohibition against directed donation of fetal tissue
- A separation between tissue procurement personnel and abortion clinic personnel
- A prohibition against any alteration of the timing of or procedures used in an abortion solely for the purpose of obtaining tissue[2]

Those involved in research uses of stem cells derived from fetal tissue would be indirectly responsible for abortions if the perceived benefits (or promise of benefits) of the research contributed to an increase in the number of abortions. Opponents of fetal tissue research argue that it is not realistic to suppose that we can always keep a woman's decision to abort separate from considerations of donating fetal tissue, as many women facing the abortion decision are likely to have already gained knowledge about fetal tissue research through widespread media attention to the issue. The knowledge that having an abortion might promote the common goodwill, opponents argue, tips the balance in favor of going through with an abortion for some of the women who are ambivalent about it. More generally, some also argue that the benefits achieved through the routine use of fetal tissue will further legitimize abortion and result in more socially permissive attitudes and policies concerning abortion.

Although there has been one empirical study examining whether the potential for fetal tissue transplantation is likely to influence abortion decisions, the issue remains largely speculative. The study involved a survey which asked, "If you became pregnant and knew that tissue from the fetus could be used to help someone suffering from Parkinson's disease, would you be more likely to have an abortion?"[3] Twelve percent of the women responded in the affirmative. The authors conclude that the option to donate tissue may influence some women's abortion decisions. The main deficiency of the study, however, is its reliance on a hypothetical that is stripped of the complexities of the actual circumstances a pregnant woman considering abortion might be operating under. As Dorothy Vawter and Karen Gervais have noted: "[G]iven how situation-specific women's abortion decisions are, it is unclear what useful information can be obtained from asking women global hypothetical questions about whether they believe the option to donate would affect their decision to terminate a 'generic' pregnancy sometime in the future."[4]

It is difficult to deny that there is a risk that knowledge of the promise of research on stem cells derived from fetal tissue will play a role in some abortion decisions, even if only very rarely. However, it is not clear J-5 that much moral weight ultimately attaches to this fact. One might be justified in some instances in asserting that "but for" research using fetal tissue a particular woman would not have chosen abortion. But one might assign this kind of causal responsibility to a number of factors which figure into abortion decisions without making ascriptions of complicity. For example, a woman might choose to have an abortion principally because she does not want to slow the advancement of her education and career. She might not have had an abortion in the absence of policies that encourage career development. Yet, we would not think it appropriate to charge those who promote such policies as complicit in her abortion. In both this case and that of research, the risk of abortion is an unintended consequence of a legitimate social policy. The burden on those seeking to end such policies is to show that the risks of harm resulting from the policies outweigh the benefits.[5] This minimally requires evidence of a high probability of a large number of abortions that would not have occurred in the absence of those policies. There is, however, no such evidence at present; nor is there any reason to think that it is forthcoming.

Symbolic Association

Agents can be complicit with wrongful acts for which they are not causally or morally responsible. One such form of complicity arises from an association with wrongdoing that symbolizes an acquiescence in the wrongdoing. As James Burtchaell characterizes it, "It is the sort of association which implies and engenders approbation that creates moral complicity. This situation is detectable when the associate's ability to condemn the activity atrophies."[6] Burtchaell maintains that those involved in research on fetal tissue enter a symbolic alliance with the practice of abortion in benefiting from it.

A common response to this position is that there are numerous circumstances in which persons benefit from immoral acts without tacitly approving of those acts. For example, transplant surgeons and patients may benefit from deaths resulting from murder and drunken driving but nevertheless condemn the wrongful acts.[7] A researcher who benefits from an aborted fetus need not sanction the act of abortion any more than the transplant surgeon who uses the organs of a murder victim sanctions the homicidal act.

This response has not, however, been satisfactory to opponents of fetal tissue research. They maintain that fetal tissue research implicates those involved in a different and far greater evil than is the transplant surgeon in the example above. Unlike drunken driving and murder, abortion is an institutionalized practice in which a certain class of humans (which pro-lifers regard as the moral equivalent of persons) are allowed to be killed. In this respect, some foes of abortion suggest that fetal tissue research is more analogous to research which benefits from victims of the Holocaust.[8]

But whatever one thinks of comparisons between the victims of Nazi crimes and aborted fetuses—and many are understandably outraged by them—one could concede the comparisons without concluding that fetal tissue research is morally problematic. There are, of course, some who believe that those who use data derived from Nazi experiments are morally complicit with those crimes. For example, Seidelman writes: "By giving value to (Nazi) research we are, by implication, supporting Himmler's philosophy that the subjects' lives were 'useless.' This is to argue that, by accepting data derived from their misery we are, *post mortem*, deriving utility from otherwise 'useless' life. Science could thus stand accused of giving greater value to knowledge than to human life itself."[9] But one need not adopt this stance. Instead, one can reasonably hold that the symbolic meaning of scientists' actions must be divined solely from their intentions. As Benjamin Freedman argues:

> A moral universe such as our own must, I think, rely on the authors of their own actions to be primarily responsible for attaching symbolic significance to those actions. . . . [I]n using the Nazi data, physicians and scientists are acting pursuant to their own moral commitment to aid patients and to advance science in the interest of humankind. The use of data is predicated upon that duty, and it is in seeking to fulfill that duty that the symbolic significance of the action must be found.[10]

One could likewise maintain that the symbolic significance of support for research using stem cells derived from fetal tissue lies in the desire to promote public health and save lives. This research is allied with a noble cause, and any taint that might attach from the source of the stem cells arguably diminishes in proportion to the potential good that the research may yield.

One who opposes the use of fetal tissue might reply, however, that those who use Nazi data can symbolically disassociate themselves from the immoral acts that produced the data only because the Nazi regime no longer exists. Those who use fetal tissue cannot—so the argument runs—divorce themselves from the immoral act that produces it because elective abortion is legally protected and largely socially accepted in our society. Opponents of fetal tissue research might seek to generate different intuitions about the moral permissibility of research using fetal tissue by offering a different analogy. Imagine the following: You live in a society where members of a particular racial or ethnic group are regularly killed when they are an unwanted burden. The practice is legally protected and generally accepted. Suppose that biological materials obtained from these individuals subsequent to their deaths are made available for research uses. Would it be morally problematic for researchers to use these materials?

It is likely that many who consider it permissible to use Nazi data would think it problematic to use materials derived from the deceased individuals in the hypothetical scenario. Arguably, what underlies the judgment that the research in the hypothetical is morally problematic is the belief that there is a heightened need to act in ways that protest moral wrongs where those wrongs are currently socially accepted and institutionalized. Attempts to benefit from the moral wrong in these circumstances may be viewed as incompatible with mounting a proper protest.

On the pro-life view, the hypothetical case is analogous to the case of research uses of fetal material. Hence, if we concede (at least for the sake of argument) that the fetus has the moral status of a person, it is not clear that an investigator can engage in fetal tissue research without implicitly sanctioning abortion. However, it is not obvious that this conclusion cuts against a public policy which supports fetal tissue research. Federal funding of research may be problematic where taxpayers who oppose the research would regard themselves as complicit with abortion as a result of the funding. There would, for example, be a basis for concern if fetal tissue research played an important causal role in an increase in the number of elective abortions. But funding research in which investigators are complicit with abortion through their symbolic association with it does not render opponents complicit with abortion. That a person's tax dollars are being used to support fetal tissue research does not imply that he or she acquiesces in abortion, nor does it diminish his or her ability to condemn abortion.

Consent

Many people hold that women should not be allowed to terminate a pregnancy solely for the purpose of donating fetal material. There are two consent requirements that can help insulate the decision to donate from the decision to abort: (1) informed consent for an abortion must be provided prior to an informed consent to donate the fetal tissue, and (2) in the consent process for abortion, there must be no (unsolicited) mention of the possibility of using fetal materials in research and transplantation. The most serious charge against these restrictions on the consent process is that it is disrespectful of the autonomy of women considering abortion to withhold information from them regarding the donation of fetal tissue. Because this information might be important to a woman's abortion decision, the failure to disclose the information would render the consent for the abortion ethically invalid.[11]

There are, however, a number of difficulties with this argument. First, it is not clear that information about the possibility of donation is materially relevant to the abortion decision, since, as discussed above, there is not adequate evidence that the option of donation (where financial incentives and directed donations are prohibited) would ever function as a reason for a woman to abort a fetus. Second, assuming the possibility of donation is materially relevant to some women's abortion decisions, there is an obligation not to disclose the option if it is unethical for women to abort for this purpose. Finally, if clinic personnel are permitted to discuss donation prior to obtaining a woman's consent for abortion, women may be (or feel) pressured to have an abortion, in which case the voluntariness of the consent will be in doubt.[12]

Another problem about consent concerns the matter of who has the moral authority to consent to donate fetal tissue. Some object that, from an ethical standpoint, a woman who chooses abortion forfeits her right to determine the disposition of the dead fetus. Burtchaell, for instance, argues that "the decision to abort, made by the mother, is an act of such violent abandonment of the maternal trusteeship that no further exercise of such responsibility is admissible."[13]

John Robertson argues that this position mistakenly assumes that the disposer of cadaveric remains acts as the guardian or proxy of the deceased. Instead, "a more accurate account of their role is to guard their own feelings and interests in assuring that the remains of kin are treated respectfully."[14] But even if we suppose that a woman does forfeit her moral authority to determine the disposition of her aborted fetus, it is not clear that informed consent is always ethically required for the use of cadaveric remains. The requirement of informed consent in medical practice is largely meant to protect the autonomy of persons. In the absence of any person whose autonomy must be respected, it does not seem that the failure to obtain consent violates anyone's rights.[15]

RESEARCH WITH STEM CELLS DERIVED FROM EMBRYOS

Research with stem cells obtained from human embryos poses moral difficulties that do not arise in the case of fetal tissue. Whereas researchers using fetal tissue are not responsible for the death of the fetus, researchers using stem cells derived from embryos will often be implicated in the destruction of the embryo. Researchers using stem cells derived from embryos are clearly implicated in the destruction of the embryo where they (1) derive the cells themselves, or (2) enlist others to derive the cells. However, there may be circumstances in which opponents of embryo research could not properly deem researchers who use embryonic stem cells complicit with the destruction of embryos. Suppose, for example, that X creates a cell line for his own study and later makes an unsolicited offer to share the cell line with Y so that Y may pursue her own research. Is Y implicated in X's act of destroying the embryo from which the cell line was derived? It does not seem that Y is implicated in X's act. As Robertson argues, it does not appear one can assign causal or moral responsibility for the destruction of an embryo to an investigator where his or her "research plans or actions had no effect on whether the original immoral derivation occurred."[16] Nonetheless, it does seem evident that much research with embryonic stem cells will be morally linked to the derivation of the cells (and the resulting destruction of the embryo), especially in the early stages of the research. Thus, an analysis of the ethics of research with embryonic stem cells, as well as the ethics of the funding of this research, must address the issue of the moral permissibility of destroying embryos.

The Moral Status of Embryos

The moral permissibility of destroying embryos turns principally on the moral status of the embryo. The debate about the moral status of embryos has traditionally revolved around the question of whether the embryo has the same moral status as children and adult humans, with a right to life that cannot be sacrificed for the benefit of society. At one end of the spectrum of positions is the view that the embryo is a mere cluster of cells which has no more moral standing than any other human cells. From this perspective, there are few, if any, limitations on research uses of embryos. At the other end of the spectrum is the view that embryos have the moral status of persons. On this view, research involving the destruction of embryos is absolutely prohibited. An intermediate position is that the embryo merits respect as human life, but not the level of respect accorded persons. Whether research using embryos is acceptable on this account depends upon just how much respect the embryo is thought to deserve.

While the moral permissibility of research using embryonic stem cells turns upon the status of the embryo, the prospects of mediating the standoff between opposing views on the matter are dim. A brief consideration of the competing positions will reveal some of the difficulties of resolving the issue.

The standard move made by those who deny the personhood of embryos is to identify one or more psychological or cognitive capacities that are thought essential to personhood (and a concomitant right to life) but which embryos lack. The capacities most commonly cited include consciousness, self-consciousness, and reasoning.[17] The problem faced by such accounts is that they seem either under- or overinclusive, depending on which capacities are invoked. If one requires self-consciousness or reasoning, most early infants will not satisfy the conditions for personhood. If sentience is regarded as the touchstone of the right to life, then nonhuman animals will also possess this right. Since most of those who reject the personhood of the embryo believe that newborn infants do possess a right to life and animals do not, these capacities cannot generally be accepted as morally distinguishing embryos from other human beings.

Of course, those who reject that embryos have the standing of persons can maintain that the embryo is simply too nascent a form of human life to merit the kind of respect that we accord more developed humans. However, in the absence of an account which decisively identifies the first stage of human development at which destroying human life is morally wrong, one can hold that it is not permissible to destroy embryos.

The fundamental argument of those who oppose the destruction of human embryos is that these embryos are human beings, and as such, have a right to life. The humanity of the embryo is thus thought to confer the status of a person upon it. The problem is that for some, the premise that all human beings have a right to life (i.e., are persons in the moral sense) is not self-evidently true. Indeed, some believe that the premise conflates two categories of "human beings"—namely, beings which belong to the species *homo sapiens*, and beings which belong to the moral community.[18] According to this view, the fact that a being belongs to the species *homo sapiens* is not sufficient to confer on it membership in the moral community. While it is not clear that those who advance this position can establish the point at which human beings first acquire the moral status of persons, those who oppose the destruction of embryos likewise fail to establish that we should ascribe the status of persons to human embryos.

Those constructing public policy on the use of embryos in research would do well to avoid attempting to settle the debate over the moral status of embryos. Ideally, public policy recommendations should be formulated in terms which individuals with opposing views on the status of the embryo can accept. As Thomas Nagel argues, "In a democracy, the aim of procedures of decision should be to secure results that can be acknowledged as legitimate by as wide a portion of the citizenry as possible."[19] Amy Gutmann and Dennis Thompson similarly argue that the construction of public policy on morally controversial matters should involve a "search for significant points of convergence between one's own understanding and those of citizens whose positions, taken in their more comprehensive forms, one must reject."[20]

Locating Convergence

R. Alta Charo suggests an approach for informing policy in this area that seeks to accommodate the interests of individuals who hold conflicting views on the status of the embryo. Charo argues that the issue of moral status can be avoided altogether by addressing the proper limits of embryo research in terms of political philosophy rather than moral philosophy:

> The political analysis entails a change in focus, away from the embryo and the research and toward an ethical balance between the interests of those who oppose destroying embryos in research and those who stand to benefit from the research findings. Thus, the deeper the degree of offense to opponents and the weaker the opportunity for resorting to the political system to impose their vision, the more compelling the benefits must be to justify the funding.[21]

In Charo's view, once we recognize that the substantive conflict among fundamental values surrounding embryo research cannot be resolved in a manner that is satisfactory to all sides, the most promising move is to perform some type of cost-benefit analysis in considering whether to proceed with the research. Thus, one could acknowledge that embryo research will deeply offend many people, but argue that the potential health benefits for this and future generations outweigh the pain experienced by opponents of the research.

It is, however, questionable whether Charo's analysis successfully brackets the moral status issue. One might object that placing the lives of embryos in this kind of utilitarian calculus will only seem appropriate to those who already presuppose that embryos do not have the status of persons. After all, we would expect most of those who believe—or who genuinely allow for the possibility—that embryos have the status of persons, to regard such consequentialist grounds for sacrificing embryos as problematic.

An acceptable political approach must seek to develop public policy around points of convergence in the moral positions of those who disagree about the status of the embryo. Of course, as long as the disagreement is cast strictly as one between those who think the embryo is a person with a right to life and those who think it has little or no moral standing, the quest for convergence will be an elusive one. But there are grounds for supposing that this is a misleading depiction of the conflict. Once this is recognized, it will become clear that there may be sufficient consensus on the status of embryos to justify some research uses of stem cells derived from them.

In his discussion of the abortion debate, Ronald Dworkin maintains that, despite their rhetoric, a large faction of the opposition to abortion does not actually believe that the fetus is a person with a right to life. This is revealed, he claims, through a consideration of the exceptions that they permit to their proposed prohibitions on abortion:

> It is a very common view, for example, that abortion should be permitted when necessary to save the mother's life. Yet this exception is also inconsistent with any belief that a fetus is a person with a right to live. Some people say that in this case a mother is justified in aborting a fetus as a matter of self-defense; but any safe abortion is carried out by someone else—a doctor—and very few people believe that it is morally justifiable for a third party, even a doctor, to kill one innocent person to save another.[22]

It can further be argued that one who regards the fetus as a person would seem to have no reason to privilege the mother's life over the fetus's life. Indeed, inasmuch as the mother typically bears some responsibility for putting the fetus in the position in which it needs her aid, one who regards the fetus as a person might well hold that the fetus's life should be privileged over the mother's life.

Some abortion conservatives further hold that abortion is morally permissible when a pregnancy is the result of rape or incest. Yet, as Dworkin comments, "it would be contradictory to insist that a fetus has a right to live that is strong enough to justify prohibiting abortion even when childbirth would ruin a mother's or a family's life, but that ceases to exist when the pregnancy is the result of a sexual crime of which the fetus is, of course, wholly innocent."[23]

The importance of these exceptions in the context of research uses of embryos is that they suggest we can identify some common ground between liberal and conservative views on the permissibility of destroying embryos. Conservatives who allow these exceptions implicitly hold with liberals that very early forms of human life can sometimes be sacrificed to promote the interests of other humans. While liberals and conservatives disagree about the range of ends for which embryonic or fetal life can ethically be sacrificed, they should be able to reach some consensus. Conservatives who accept that killing a fetus is permissible where it is necessary to save a pregnant woman or spare a rape victim additional trauma could agree with liberals that it is also permissible to destroy embryos where it is necessary to save people or prevent extreme suffering. Of course, the cases are different inasmuch as the existence of the fetus directly conflicts with the pregnant woman's interests, while a particular *ex utero* embryo does not threaten anyone's interests. But this distinction does not bear substantial moral weight. It is the implicit attribution of greater value to the interests of adult humans over any interests the fetus may have that informs the judgment that it is permissible to kill it in the cases at issue. Thus, the following would seem a reasonable formulation of a position on embryo research that liberals and conservatives could agree upon: *Research that involves the destruction of embryos is permissible where there is good reason to believe that it is necessary to cure life-threatening or severely debilitating diseases.*

Given the great promise of stem cell research for saving lives and alleviating suffering, this position would appear to permit research uses of stem cells derived from embryos. Some might object, however, that the benefits of the research are too uncertain to justify a comparison with the abortion exceptions. But the lower probability of benefits from research uses of embryos is balanced by a much higher ratio of potential lives saved for lives lost. Another objection is that it is unnecessary to use embryos for stem cell research because there are alternative means of obtaining stem cells. This reflects an important concern. The derivation of stem cells from embryos is justifiable only if there are no less morally problematic alternatives for advancing the research. At present, there appear to be strong scientific reasons for using embryos. But this is a matter that must continually be revisited as science advances.

It is important to further note that the abortion exceptions that serve as the basis for the type of convergence identified above are exceptions to the law banning federal funding for abortions.[24] Thus, federal funding of research uses of embryos within such limitations as those specified above can reasonably be viewed as consistent with current federal funding practices in the abortion context.

There is, however, one further qualification that must be made here. The justification developed above for federal funding of research that involves the destruction of embryos does not appear to extend to research in which embryos are created expressly for research purposes. While there is wide agreement that it is acceptable to produce embryos for *in vitro* fertilization (IVF), there is much controversy about the moral permissibility of creating embryos for research. Indeed, many who approve of research using embryos remaining from IVF treatments oppose the creation of embryos solely for research purposes.[25] The main basis for the distinction is that, whereas embryos created for procreative purposes are originally viewed as potential children, embryos created for research are meant to be treated as mere objects of study from the outset. Some dispute that there is a morally relevant distinction between the use of spare IVF embryos and embryos created for research.[26] But, regardless of which view is more sound, the foundation for agreement on embryo research identified above does not appear to support the use of embryos that are created expressly for research purposes. This limitation on federal funding of stem cell research could ultimately prove to be a significant impediment to the advancement of the research. However, it should come as little surprise that parties on both sides of the issue may incur substantial costs when constructing public policy around points of convergence between competing visions of the meaning and value of life.

NOTES

1. Letter to Congress and President Clinton, March 4, 1999.

2. National Institutes of Health (NIH), *Report of the Fetal Tissue Transplantation Research Panel* (Bethesda, Md.: NIH, 1988).

3. D. K. Martin, et al., "Fetal Tissue Transplantation and Abortion Decisions: A Survey of Urban Women," *Canadian Medical Association Journal* 153, no. 5 (1995): 548.

4. D. E. Vawter and K. G. Gervais, "Commentary on Abortion and Fetal Tissue Transplantation," *IRB: A Review of Human Subjects Research* 15, no. 3 (1993): 5.

5. J. F. Childress, "Ethics, Public Policy, and Human Fetal Tissue Transplantation Research," *Kennedy Institute of Ethics Journal* 1, no. 2 (1991): 93–121.

6. J. T. Burtchaell, "The Use of Aborted Fetal Tissue in Research: A Rebuttal," *IRB: A Review of Human Subjects Research* 11, no. 2 (1989): 9–12.

7. J. Robertson, "Fetal Tissue Transplant Research Is Ethical," *IRB: A Review of Human Subjects Research* 6, no. 10 (1988): 5–8; D. Vawter et al., "The Use of Human Fetal Tissue: Scientific, Ethical, and Policy Concerns," *International Journal of Bioethics* 2, no. 3 (1991): 189–96. Dorothy Vawter et al., *The Use of Human Fetal Tissue: Scientific, Ethical, and Policy Concerns* (Minneapolis: Center for Biomedical Ethics, University of Minnesota, 1990).

8. J. Bopp Jr., "Fetal Tissue Transplantation and Moral Complicity with Induced Abortion" in *The Fetal Tissue Issue: Medical and Ethical Aspects*, ed. P. Cataldo and A. Moraczewski (Braintree, Mass.: Pope John Center, 1994), pp. 6–-79.

9. W. E. Seidelman, "Mengele Medicus: Medicine's Nazi Heritage," *Milbank Quarterly* 66 (1988): 221–39.

10. B. Freedman, "Moral Analysis and the Use of Nazi Experimental Results" in *When Medicine Went Mad*, ed. A. L. Caplan (Totowa, N.J.: Humana Press, 1992), p. 151.

11. D. K. Martin, "Abortion and Fetal Tissue Transplantation," *IRB: A Review of Human Subjects Research* 15, no. 3 (1993): 1–3.

12. Vawter and Gervais, "Commentary on Abortion and Fetal Tissue Transplantation."

13. Burtchaell, "The Use of Aborted Fetal Tissue in Research: A Rebuttal."

14. Robertson, "Fetal Tissue Transplant Research Is Ethical," p. 6.

15. D. G. Jones, "Fetal Neural Transplantation: Placing the Ethical Debate within the Context of Society's Use of Human Material," *Bioethics* 3, no. 1 (1991): 22–43.

16. J. Robertson, "Ethics and Policy in Embryonic Stem Cell Research," *Kennedy Institute of Ethics Journal* 2, no. 9 (1999): 109–36.

17. M. A. Warren, "On the Moral and Legal Status of Abortion," *Monist* 57 (1973): 43–61; M. Tooley, *Abortion and Infanticide* (New York: Oxford University Press, 1983); J. Feinberg, "Abortion," in *Matters of Life and Death*, ed. T. Regan (New York: Random House, 1986), pp. 256–93.

18. Warren, "On the Moral and Legal Status of Abortion."

19. T. Nagel, "Moral Epistemology," in *Society's Choices*, ed. R. E. Bulger, E. M. Bobby, and H. V. Fineberg (Washington, D.C.: National Academy Press, 1995), pp. 201–14.

20. A. Gutmann and D. Thompson, *Democracy and Disagreement* (Cambridge, Mass.: Belknap Press, 1996).

21. R. A. Charo, "The Hunting of the Snark: The Moral Status of Embryos, Right-to-Lifers, and Third World Women," *Stanford Law and Policy Review* 6 (1995): 11–27.

22. R. Dworkin, *Life's Dominion: An Argument about Abortion, Euthanasia, and Individual Freedom* (New York: Vintage, 1994), p. 32.

23. Ibid.

24. Senate Appropriations Subcommittee on Labor, Health and Human Services, and Education, *Title V*, 112 Stat. 3681-385, Sec. 509(a) (1) and (2).

25. G. Annas, A. L. Caplan, and S. Elias, "The Politics of Human-Embryo Research: Avoiding Ethical Gridlock," *New England Journal of Medicine* 334, no. 20 (1996.): 1329–32.

26. J. Harris, *Clones, Gene, and Immortality: Ethics and the Genetic Revolution* (New York: Oxford University Press, 1998).

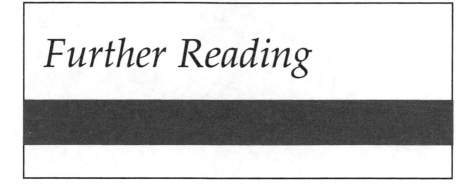

Further Reading

As you will have realized, stem cell research is a rapidly forward moving field, scientifically, medically, and politically. To keep yourself abreast, you should be reading at least one of the major newspapers like the *New York Times* or the *Washington Post* (whose treatments of the topic tend to be truly excellent), as well as looking at some of the more popular science magazines like the *Scientific American* and (in Britain) the *New Scientist*. The two weekly, more professional science magazines, *Science* in America and *Nature* in Britain, also carry much good and pertinent material. A good collection of rather technical papers dealing with the science is *Stem Cell Biology*, edited by Daniel R. Marshak, Richard L. Gardner, and David Gottlieb (Cold Spring Harbor Laboratory, 2001), and an excellent collection of papers dealing with more general issues is *The Human Embryonic Stem Cell Debate (Basic Bioethics)*, edited by Suzanne Holland, Karen Lebacqz, and Laurie Zoloth (MIT Press, 2001). You might also want to look at two earlier collections published by Prometheus Books in the same series as this dealing with related scientific, moral, and religious issues. These are *Cloning: Responsible Science or Technomadness?* edited by Michael Ruse and Aryne Sheppard (Prometheus Books, 2000), and *Genetically Modified Foods: Debating Biotechnology*, edited by Michael Ruse and David Castle (Prometheus Books, 2002).

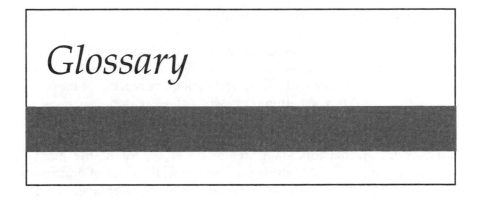

Glossary

Definitions with a "*" are found in the National Bioethics Advisory Commission report, *Ethical Issues in Human Stem Cell Research*, September 1999.

adult stem (AS) cells:* stem cells found in the adult organism (e.g., bone marrow, skin, intestine) that replenish tissues in what cells often have limited life spans. They are more differentiated than embryonic stem (ES) cells or embryonic germ (EG) cells.

a priori: a claim as in logic and pure mathematics, the truth of which is not dependent on experiment or sense experience.

amino acids: the building blocks of proteins.

antibodies: proteins that bind to foreign chemicals or microorganisms (antigens) to make them harmless; an important component of the immune system.

ART (assisted reproductive technology):* all treatments or procedures that involve handling human eggs and sperm for the purpose of helping a woman become pregnant. Types of ART include *in vitro* fertilization, gamete infrafallopian transfer, zygote intrafallopian transfer, embryo cryopreservation, egg or embryo donation, and surrogate birth.

artificial insemination (AI): a process by which sperm are collected from males and deposited in the females by instruments.

artificial selection: selection by humans and breeding of those individual plants or animals that possess desirable traits.

asexual reproduction: reproduction where there is no fusion of gametes and in which the genetic makeup of parents and off-spring is usually identical.

bioethics: a discipline concerned with the application of ethics to biological problems, especially in the field of medicine.

bioreactors: a containment vessel for biological reactions, used in particular for fermentation processes and enzymatic reactions.

biotechnology: technological processes (agriculture, manufacturing, medicine) that involve the use of biological systems.

blastocyst:* the blastula stage in the development of the mammalian embryo (* a mammalian embryo in the state of development that follows the morula. It consists of an outer layer or trophoblast to which is attached an inner cell mass).

blastomere:* one cell of the blastula; a cell formed during early cleavage of a vertebrate embryo (* one of the cells into which the egg divides after it is fertilized; one of the cells resulting from the division of a fertilized ovum).

blastomere separation: the process in which the cells in blastomeres are made to separate; cloning through blastomere separation yields embryos with identical sets of nuclear DNA *and* mitochondrial DNA; also called embryo splitting.

blastula: usually a spherical structure produced by cleavage of a fertilized egg cell; consists of a single layer of cells surrounding a fluid-filled cavity called the blastocoel.

blood plasma: the fluid portion of blood containing proteins, salts, and other substances, but excluding the blood cells.

carcinogen: a chemical that causes cancer, generally by altering the structure of DNA.

cell cleavage: first of several cell divisions in early embryonic development that converts the zygote into a multicellular embryo.

cell culture: growth of tissue cells in artificial media.

cell cycle: the series of events in the life of a cell consisting of mitosis or nuclear division, cytokenesis or cytoplasmic division, and the interphase during which typically cells grow.

cell division: the process by which cells multiply.

cellular differentiation: see **differentiation**.

chimera:* on organism composed of two genetically distinct types of cells.

chromatin: the complex of DNA, protein, and RNA, that makes up chromosomes.

chromosome: a subcellular structure containing DNA plus the proteins that are associated with the DNA.

clone: multiple copies of identical DNA sequences that are produced when they are inserted into cloning vehicles (plasmids and other vectors) and replicated in bacteria using these cloning vehicles.

cloning:* the production of a number of genetically identical DNA molecules by inserting the chosen DNA into a phage or plasmid vector; the vector then being used to infect a suitable host within which the chosen DNA is replicated to form a number of copies, or clones (* production of a precise genetic copy of a molecule [including DNA], cell, tissue, plant, or animal).

conjoined twins: offspring that are physically attached at birth; occurs when the early embryo subdivides forming two groups of cells that do not separate completely; also referred to as Siamese twins.

consequentialist ethics: morally right actions are those that maximize the good. See **utilitarianism**.

cytoplasm: general cellular contents exclusive of the nucleus.

dedifferentiation: the opposite process of differentiation where cells or structures revert back to a less specialized stage.

deontological ethics: a moral theory according to which the rightness or necessity of an action is not exclusively determined by the value of its consequences.

developmental biology: the study of the processes by which cells and cell structures typically become more complex.

differentiated cell: a cell that has undergone the process of cellular differentiation and typically becomes specialized, e.g., a muscle cell.

differentiation:* a process changing a young, relatively unspecialized cell to a more specialized cell (* specialization of characteristics or functions of cell types).

diploid:* the condition of having a paired set of chromosomes (* cell containing two complete sets of genes derived from the father and mother, respectively; normal chromosome complement of somatic cells; in humans, forty-six chromosomes).

dizygotic twins: offspring that develop when two eggs are fertilized by a different sperm; also called fraternal twins.

DNA (deoxyribonucleic acid): a long, chainlike molecule that transmits genetic information.

E. coli (*Escherichia coli*): a bacterium whose natural habitat is the gut of humans and other warm-blooded animals.

ectoderm:* the outer of the three tissue layers of the early animal embryo; typically gives rise to the skin and nervous systems; see also **mesoderm** and **endoderm** (* outer layer of cells in the embryo; origin of skin, pituitary gland, mammary glands, and all parts of the nervous system).

embryo:* a young multicellular organism that develops from the zygote and has not yet become free-living; the developing human organism until the end of the second month, after which it is referred to as a fetus (* [1] beginning of any organism in the early stages of development, [2] a state (between ovum and fetus) in the prenatal development of a mammal, [3] in humans, the state of development between the second and eighth weeks after fertilization, inclusive).

embryo splitting: see **blastomere separation.**

embryology: the scientific study of the development of organisms during the embryonic stage.

embryonic stem (ES) cell: a formative cell capable of giving rise to all other kinds of cells.

embryonic stem (ES) cells:* cells that are derived from the inner cell mass of a blastocyst embryo.

embryonic germ (EG) cells:* cells that are derived from precursors of germ cells from a fetus.

end-in-itself: a Kantian concept that something has value independent of external material ends.

endoderm:* the inner layer of the three basic tissue layers that appear during early development; becomes the digestive tract and its outgrowths—the liver, lungs, and pancreas; see also **ectoderm** and **mesoderm** (* innermost of the three primary layers of the embryo; origins of the digestive tract, liver, pancreas, and lining of the lungs).

enucleated cell or egg: a cell or egg where the nucleus has been removed.

enucleation: the process of removing the nucleus from a cell or egg.

enzyme: a protein that speeds or slows a specific chemical reaction.

epigenesis: the theory that an embryo develops from a structureless cell by the gradual addition of new parts.

epigenetic change: any change of gene activity during the development of the organism from the fertilized egg to the adult.

epithelial cells: cells that make up the tissue that covers body surfaces, lines body cavities, and forms glands.

eugenics: the claim that the way to improve humankind is through selective breeding.

ex utero:* outside of the uterus.

fertility: the state or capacity for producing offspring.

fertilization: the fusion of male and female gametes resulting in the formation of a zygote.

fetus, fetal: an unborn offspring in a late stage of development; from the third month of pregnancy to birth in humans.

fibroblasts:* connective tissue cells that secrete the intercellular material of, for example, bones and cartilage (* a cell present in connective tissue, capable of forming collagen fibers).

gamete:* a sex cell; in plants and animals, an egg or sperm (* [1] any germ cell, whether ovum or spermatozoon, [2] a mature male or female reproductive cell).

gastrula: early stage of development that follows the blastula stage.

gtastrulation:* process of transforming the blastula into the gastrula, at which point embryonic germ layers or structures begin to be laid out.

gene: the unit of heredity that consists of a stretch of DNA; contains information for the construction of other molecules, mostly proteins.

gene therapy: the use of DNA sequences either to switch off defective genes or to insert genes that specify the correct product.

genetic determinism: the claim that the traits of organisms are a direct function of their genes, not subject to environmental influences.

genetic diversity: a measure of the differences in allelic frequencies between isolated or semi-isolated populations due to various evolutionary forces such as selection and genetic drift.

genetic engineering: the manipulation of the DNA content of an organism to alter the characteristics of that organism.

genetic modification (GM): a process (especially an artificial process) that results in a change in the genetic makeup of a population.

genetic selection: artificial selection of specific genes.

genome: the genetic material of an organism.

genotype: the organism's genetic information, as distinguished from its physical appearance (phenotype).

germ cell:* cells within the body that give rise to gametes (* gametes [ova and sperm] or cells that give rise directly to gametes).

gestation: the duration of pregnancy.

G0 phase: phase in the mitotic cycle; typically a stage of arrested cellular activity.

growth phase: see **interphase**.

haploid:* the condition of having a single set of chromosomes, received from father or mother (* a cell with half the number of chromosomes as the somatic diploid cell, such as the ova or sperm. In humans, the haploid cell contains twenty-three chromosomes).

histocompatibility antigens: systems of allelic alloantigens that can stimulate an immune response leading to transplant rejection when the donor and recipient are mismatched.

hormones: "messenger" molecules that coordinate the actions of various tissues of the body; hormones produce a specific effect on the activity of target cells remote from their point of origin.

identical twins: see **monozygotic twins**.

immunology: the study of specific defense mechanisms (immune responses).

implantation: the attachment of the developing embryo to the uterine wall.

infertility: the inability of a couple to achieve conception (fertilization and establishment of pregnancy) through sexual intercourse

interphase: phase of the cell cycle during which growth occurs.

intracytoplasmic sperm injection: the process by which sperm cells are inserted directly into an egg cell.

in utero: occurring in the uterus.

in vitro:* occurring outside a living organism (literally "in glass") (* in an artificial environment, such as a test tube or culture medium).

in vivo:* occurring in a living organism (* in the natural environment, i.e., within the body).

IVF (in vitro fertilization):* the process during which an egg is removed from the ovary and fertilized with sperm in laboratory glassware and then reimplanted in the female (* a process by which a woman's eggs are extracted and fertilized in the laboratory and transferred after they reach the embryonic stage into the woman's uterus through the cervix. Roughly 70 percent of assisted reproduction attempts involve IVF using fresh embryos developed from a woman's own eggs).

Kantian ethics: based on the categorical imperative, namely that we should act only according to those maxims that can be consistently willed as a universal law.

karyotype:* chromosome characteristics of an individual cell or cell line, usually presented as a systematic array of metaphase chromosomes from a photograph of a single cell nucleus arranged in pairs in descending order of size.

keratinocyte: a specialized epidermal cell that synthesizes keratin, the substance that makes up nails and feathers.

lactation: the production or release of milk from the breast.

mammary gland: the organs that secrete milk to feed offspring.

meiosis: process in which a diploid cell undergoes two successive nuclear divisions to produce four haploid cells; the process that produces gametes in animals.

mesoderm:* the middle layer of the three basic tissue layers that develop in the early embryo; gives rise to connective tissue, muscle, bone, blood vessels, and many other structures; see also **ectoderm** and **endoderm** (* middle of the three primary germ layers of the embryo; origin of all connective tissues, all body musculature, blood, cardiovascular and lymphatic systems, most of the urogenital system, and lining of the pericardial, pleural, and peritoneal cavities).

metabolism: a collective term for the chemical processes involved in the maintenance of life and by which energy is made available.

metaphase: the stage of mitosis and meiosis during which the chromosomes line up on the equatorial plane of the cell.

metaphysics: the branch of philosophy which deals with the ultimate nature and origin of things.

microinjection: injection of cells with solutions by using a micropipette.

miscarriage: the expulsion of the embryo or fetus from the uterus before it is viable.

mitochondria (sing. **mitochondrion**): a specialized subcellular structure that converts chemical energy from one form to another.

mitochondrial disease: diseases that are due to mutations or malfunctions in mitochondria.

mitochondrial DNA (mtDNA): the DNA contained in the mitochondria

mitosis: division of the cell nucleus, resulting in two daughter
nuclei with the same number of chromosomes as the parent
nucleus; consists of four phases: prophase, metaphase,
anaphase, and telophase.

molecular biology: the chemistry and physics of the molecules that
constitute living things.

monoclonal antibody: an antibody produced by a clone of anti-
body-producing cells that recognizes only one feature.

monovular twins: see **monozygotic twins.**

monozygotic twins: offspring that possess identical sets of genetic
material; result from the division of the early embryo; also
referred to as identical or monovular twins.

moral agency: the capability of having rights as an individual.

morphology: the scientific study of organic form.

morula:* (1) mass of blastomeres resulting form early cleavage divisions of the zygote, (2) solid mass of cells resembling a mulberry,
resulting form cleavage of an ovum.

multipotent: see **totipotent.**

mutagen: any agent that is capable of producing changes in the
DNA.

mutations: any change in the DNA , including a change in the
sequence of nucleotide pairs that constitute DNA or a rearrange-
ment of genes within chromosomes.

mutatis mutandis: after the necessary or appropriate changes have
been made.

myoepithelial cells: epithelial cells of coelenterata that are provided
with basal contractile outgrowths.

natural selection: the key Darwinian mechanism for evolutionary
change, claiming that only a percentage of organisms in each
generation survive and reproduce, and that these do so because
they have characteristics which other members of the population do not possess; these "adaptive" characteristics are passed
along to offspring.

nuclear: pertaining to the nucleus.

nuclear DNA: the DNA contained within the nucleus of the cell (as
opposed to the DNA contained in the mitochondria).

nuclear substitution: see **nuclear transfer.**

nuclear transfer: process where the nucleus of a host cell is removed and replaced with the nucleus of a donor cell that is to be cloned

nuclear transplantation: see **nuclear transfer**.

nucleolus (pl. nucleoli): small nuclear body made up of protein and ribosomal RNA.

nucleus (pl. nuclei): a double membrane-bounded cellular structure containing DNA.

neurula: the stage of development of the vertebrate embryo that is characterized by the development of the neural tube from the neural plate.

nullipotent: unable to direct the development of cells unlike themselves.

oestrus: the recurrent period of heat or sexual receptivity occurring around ovulation in female mammals.

oncogenesis: the processes of tumor formation.

oocytes:* cells that give rise to egg cells (ova) by meiosis (* [1] diploid cell that will undergo meiosis (type of cell division of germ cells) to form an egg, [2] immature ovum).

ovulation: the release of a mature egg from the ovary.

ovum:* female reproductive or germ cell.

parthenogenesis: a form of asexual reproduction in which an unfertilized egg develops into an adult organism.

parturition: the process of birth.

pathogen: a disease-causing agent.

perinatal: surrounding the time of birth.

pharmaceuticals: chemicals used in medicine.

phenomenology: the philosophy in which supposedly we come to know mind as it is in itself through the study of the way in which it appears to us.

phenotype: the physical and behavioural characteristics of an organism.

physiology: the scientific study of the workings of living bodies and their parts.

ploidy: relating to the number of sets of chromosomes in a cell.

pluripotent:* capable of giving rise to various kinds of embryonic tissue (* cells, present in the early stages of embryo development, that can generate all cell types in a fetus and in the adult and that are capable of self-renewal. Pluripotent cells are not capable of developing into an entire organism).

polyclonal antibody: an antibody produced by many clones of cells that recognize several molecular features of an antigen.

polymorphic loci: genetic loci, in a population, which have several different alleles.

post mortem: the period after death; also, the examination of an organism after death.

pre-embryos: the preimplantation stages of the development of the zygote, occurring during the first two weeks after fertilization.

preformation theory: the theory of early developmental biologists that the fully formed animal or plant exists in a minute form in the germ cell; see **epigenesis**.

preimplantation diagnosis: the determination of the genotype of an *in vitro*–fertilized human embryo (for genetic diseases) prior to its implantation.

preimplantation embryo:* (1) embryo before it has implanted in the uterus, (2) commonly used to refer to *in vitro*–fertilized embryos before they are transferred to a woman's uterus.

prenatal diagnosis (or screening): fetal cells are sampled and analyzed for chromosomal and biochemical disorders.

prima facie: clear and apparent at first glance, presumed true unless disproven by subsequently acquired evidence.

progenitors: the ancestors of an organism.

progeny: offspring.

promoter region: a site on the DNA that facilitates the initiation of transcription.

pronucleus (pl. pronuclei): one of the two nuclear bodies of a newly fertilized egg cell; the male and the female pronuclei fuse to form the nucleus of the new cell.

quiescence: the state of dormancy or inactivity.

recessive gene: any gene whose expression is masked by another (dominant) gene.

recombinant DNA: hybrid DNA produced by joining pieces of DNA from different sources.

reductio ad absurdum: refutation of an assumption by deriving a contradiction (or otherwise clearly false conclusion) from the original assumption.

reproductive cloning: the use of cloning technology to produce new individuals (in contrast to therapeutic cloning).

senescence: the aging process.

sexual reproduction: form of reproduction in which two gametes fuse to form a zygote.

Siamese twins: see **conjoined twins**.

slippery slope: an argument that an action apparently unobjectionable in itself would set in motion a train of events leading ultimately to an undesirable outcome.

somatic cell:* body cell, as distinguished from a germ cell (* [from *soma*, the body] (1) cells of the body which in mammals and flowering plants normally are made up of two sets of chromosomes, one derived from each parent, (2) all cells of an organism with the exception of germ cells).

somatic cell nuclear transfer (SCNT): the process by which a nucleus from a (somatic) body cell is removed and inserted into an egg cell.

spermatogonium: germ cell produced by a male.

stem cell:* see **embryonic stem (ES) cell** (* cells that have the ability to divide indefinitely and to give rise to specialized cells as well as to new stem cells with identical potential).

surrogate: a female whose uterus is used to gestate an embryo that is not her own.

teleology: a theory that describes or explains in terms of purpose.

telomerase: the enzyme that elongates telomeres.

telomere: the end of a chromosome in which the DNA is looped back.

telos: endpoint, aim, goal, or purpose of an activity, or a final state of affairs.

therapeutic cloning: the use of cloning technology for medical purposes, for example, the production of organs for transplantation (in contrast to reproductive cloning).

tissue culturing: a method of propagating healthy cells outside the organism.

totipotency (adj. **totipotent**):* ability of a cell (or nucleus) to develop into any other type of cell (or nucleus) (* having unlimited capacity. Totipotent cells have the capacity to differentiate into the embryo and into extraembryonic membranes and tissues. Totipotent cells contribute to every cell type of the adult organism).

transcription: the process of converting information in DNA into information in RNA (ribonucleic acid); RNA functions mainly in protein synthesis.

transgene: a gene that is taken from one organism and inserted into the genome of another.

transgenesis: genetic modification of animals using transgenes.

transgenic animal: an animal that has incorporated foreign DNA into its genome.

translation: the process of converting the information in RNA into protein; also called protein synthesis.

trimester: one of the three three-month periods of human pregnancy

trophoblast:* outermost layer of the developing blastocyst of a mammal. It differentiates into two layers, the cytotrophoblast and syntrophoblast, the latter coming into intimate relationship with uterine endometrium with which it establishes a nutrient relationship.

undifferentiated cell: an unspecialized cell that has not undergone differentiation.

unfertilized egg: an egg cell that has not undergone fertilization.

utilitarianism: the moral theory that an action is morally right if and only if it produces at least as much good (utility) for all people affected by the action as any alternative action.

vector: an agent, such as a bacterial plasmid or virus, that transfers genetic information.

xenotransplantation: transplantation of tissue or organs between species.

zygote:* one-celled embryo formed by the fusion of two germ cells (* [1] cell resulting from fusion of two gametes in sexual reproduction, [2] fertilized egg (ovum), [3] diploid cell resulting from union of sperm and ovum, [4] developing organism during the first week after fertilization).

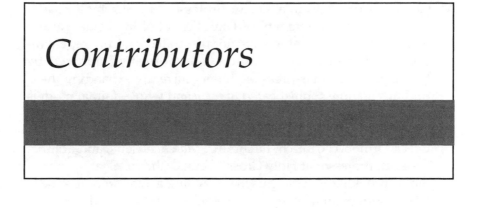

Contributors

MARCIA BARINAGA is a contributing correspondent (Berkeley Bureau) for *Science*. She writes about neuroscience and other areas of basic biology; also writes policy and other stories with a West Coast angle.

GEORGE W. BUSH is the forty-third president of the United States, formerly the forty-sixth governor of the State of Texas. He is the son of the forty-first president of the United States, George Herbert Walker Bush; and he is the second son of a president to also become president, after John Quincy Adams, son of John Adams, who was elected in 1824.

ARTHUR CAPLAN has been the director of the Center for Bioethics and trustee professor of Bioethics at the University of Pennsylvania since 1994. He is currently chairman of the advisory committee to the Department of Health and Human Services, Centers for Disease Control and Prevention, and Food and Drug Administration on blood safety and availability.

RONALD COLE-TURNER is the H. Parker Sharp Professor of Theology and Ethics at Pittsburgh Theological Seminary, a position which relates theology and ethics to developments in science and technology. He is an ordained minister of the United Church of Christ and chairs the UCC committee on genetics. He is also the author of the popular baptism hymn, "Child of Blessing, Child of Promise."

MAUREEN L. CONDIC is an assistant professor of Neurobiology and Anatomy at the University of Utah School of Medicine. Recent work from her laboratory has demonstrated that the regeneration of adult neurons in culture can be greatly improved by transgenic integrin expression. She is currently expanding these findings to more sophisticated *in vitro* and whole-animal models of adult regeneration.

FATHER DEMETRIOS DEMOPULOS is the parish priest of Holy Trinity Greek Orthodox Church, Fitchburg, Massachusetts, and adjunct assistant professor at Holy Cross Greek Orthodox School of Theology, Brookline, Massachusetts, teaching a Templeton Foundation award–winning course on science and religion. His research interests involve exploring the intersection of science and Orthodox theology. Before becoming a priest, Father Demopulos was a geneticist, with appointments at Arizona State University, University of Alabama-Birmingham, and Children's Hospital, Boston, Massachusetts. He has coauthored articles on maize and human genetics in various journals

RABBI ELLIOT N. DORFF is rector and distinguished professor of philosophy at the University of Judaism in Los Angeles. He is vice chair of the Committee on Jewish Law and Standards of the Conservative Movement in Judaism.

GERON ETHICS ADVISORY BOARD, an ethics advisory board whose members represent a variety of philosophical and theological traditions with a breadth of experience in health care ethics, was created by Geron Corporation in 1998. The board functions as an independent entity, consulting and giving advice to the corporation on ethical aspects of the work Geron sponsors. Members of the board have no financial interest in Geron Corporation.

SIDNEY HOUFF is professor and chairman of the Department of Neurology at the Loyola University Health Systems at Loyola University, Chicago. He is most recognized as a researcher who has studied AIDS-related neurological diseases and was the first scientist to report human-to-human transmission of rabies via corneal transplants. Dr. Houff plans on continuing research in how viruses alter areas of the brain responsible for behavior and cognitive function and as chairman will expand the comprehensive epilepsy program, multidisciplinary stroke center, and broaden neurology services to pediatrics. Long-term goals include further integration of clinical and basic sciences in the study of a number of neurological diseases.

ALINE H. KALBIAN is an assistant professor of religion at Florida State University. She teaches in the area of religious ethics and gender and ethics. Her areas of interest include contemporary Catholic moral theology, gender and ethics, medical ethics, and sexual ethics. She is presently working on the book *Controlling Reproduction/Reproducing Gender: The Catholic Church and Human Sexuality*. Her other research interests include ethics and reproductive technology and gender metaphors in recent Catholic magisterial writings.

KATTY KAY AND MARK HENDERSON are reporters for the *Times* in London.

JANE MAIENSCHEIN is a professor of philosophy and biology at Arizona State University in Tempe, Arizona. Dr. Maienschein specializes in the history and philosophy of biology. Looking at research in embryology, genetics, and cytology in the first half of this century, she asks what factors have brought changes in biological theory and practice. She has concluded that changing epistemological convictions about what counted as good science caused researchers to address increasingly narrower and more specialized questions and to look for concrete empirical results. Biologists began to accept that they were generating temporary working hypotheses rather than always finding the "truth," but they regarded their "facts" as definitive. As the science became more technically complex, a concept of "cutting-edge" science emerged and began to define success in biology. Dr. Maienschein combines detailed analysis of the epistemological standards, theories, laboratory practices, organisms selected, techniques, and equipment with study of the people, institutions, and their historical settings to explore the changing nature of the biological sciences and the social, political, and legal context in which science thrives.

CAROL MARIN was named a correspondent for CBS News in October 2000. In this role, Marin primarily serves as a contributor to *60 Minutes* and *60 Minutes II*.

GLENN MCGEE is an associate director at the Center for Bioethics of the University of Pennsylvania, where he holds appointments in philosophy, history, sociology of science, and cellular engineering. He is editor in chief of the *American Journal of Bioethics* and author of a number of articles about ethical issues in genetics, stem cell research, cloning, and reproduction.

NATIONAL BIOETHICS ADVISORY COMMISSION **(NBAC)** was created by President William Jefferson Clinton with Executive Order 12975 on 3 October 1995. The purpose of the NBAC was to provide a review of policies and procedures in the area of bioethics. The commission has produced reports on the subjects of cloning and stem cell research. Directed by Harold T. Shapiro, NBAC's charter expired on 3 October 2001.

NATIONAL INSTITUTES OF HEALTH (NIH) is one of eight health agencies of the Public Health Services which, in turn, is part of the U.S. Department of Health and Human Services. Comprised of twenty-seven separate components, mainly institutes and centers, NIH has seventy-five buildings on more than three hundred acres in Bethesda, Maryland. From a total of about $300 in 1887, the NIH budget grew to more than $20.3 billion in 2001.

MICHAEL NOVAK is the George Frederick Jewett Scholar in Religion, Philosophy, and Public Policy Director of Social and Political Studies at the American Enterprise Institute in Washington, D.C. He received the Templeton Prize for Progress in Religion (a million-dollar purse awarded at Buckingham Palace) in 1994, and delivered the Templeton address in Westminster Abbey. He is a frequent contributor to *National Review*, *First Things*, and *New Republic*.

DANIEL PERRY is executive director of the Alliance for Aging Research (www.agingresearch.org), a Washington, D.C.–based nonprofit organization dedicated to improving the health and independence of aging Americans through public and private funding of medical research and geriatric education.

CHRISTOPHER A. PYNES is an instructor and the undergraduate advisor for the Department of Philosophy at Florida State University. His dissertation, *Human Cloning and Moral Status*, was written under the supervision of Michael Ruse. He has published articles in logic, critical thinking, and vegetarianism in Plato's *Republic*. His philosophical interests are in ethics, biomedical ethics, logic, philosophy of science, and analytic jurisprudence.

JOHN A. ROBERTSON holds the Vinson and Elkins Chair at the University of Texas School of Law at Austin, and currently serves as cochair of the ethics committee of the American Society for Reproductive Medicine.

MICHAEL RUSE is Lucyle T. Werkmeister Professor of Philosophy at Florida State University. He began his teaching career at the University of Guelph, where he taught for thirty years. He is a fellow of the Royal Society of Canada. He writes on the nature of science, in particular evolutionary biology, and the nature of value. A list of his most recent books would include the following: *Can a Darwinian Be a Christian? The Relation between Science and Religion; Mystery of Mysteries: Is Evolution a Social Construction?;* and *Monad to Man: The Concept of Progress in Evolutionary Biology.*

KENNETH J. RYAN is the Kate Macy Ladd Distinguished Professor of Obstetrics, Gynecology, and Reproductive Biology, Emeritus at Harvard University. From 1973 to 1993, Dr. Ryan served as chairman of the Department of Obstetrics, Gynecology, and Reproductive Biology at Harvard and the Brigham and Women's Hospital. From 1974 to 1978 he served as chairman of the National Commission for the Protection of Human Subjects of Biomedical and Behavioral Research. Dr. Ryan is former chairman of the ethics committee of the Brigham & Women's Hospital and serves on many professional and governmental committees and panels related to ethics and public policy in medicine and human subjects research. He has received numerous awards and fellowships and is published in the areas of endocrinology, human reproduction, and medical ethics.

ABDULAZIZ SACHEDINA was born in Tanzania. Abdulaziz Sachedina is professor of religious studies at the University of Virginia. His Ph.D. is from the University of Toronto, and he has a B.A. from Aligarh Muslim University in India and Ferdowsi University in Iran. He has been visiting professor at Wilfrid Laurier, Waterloo, and McGill Universities in Canada, Haverford College and the University of Jordan, Amman. He has lectured widely in east Africa, India, Pakistan, Europe, and the Middle East. Professor Sachedina is a core member of the "Islamic Roots of Democratic Pluralism," project in the CSIS Preventive Diplomacy program and a key contributor to the program's efforts to link religion to universal human needs and values in the service of peacebuilding.

NATHAN SALLEY at the age of sixteen testified before Congress on 17 July 2001. He stated that as a survivor of acute myloid leukemia thanks to the use of stem cells research from cord blood, he would prefer that embryos not be used for stem cell research.

ANDREW W. SIEGEL is associate director of academic programs at the Bioethics Institute. He received his J.D. and Ph.D. in philosophy from the University of Wisconsin at Madison. He was a Green-wall Fellow at Johns Hopkins and Georgetown, 1996–1998. Dr. Siegel has served as staff philosopher for the National Bioethics Advisory Commission, legislative fellow for Senator Edward M. Kennedy and the Labor and Human Resources Committee, and staff attorney for the task force on genetic testing of the Working Group on the Ethical, Legal, and Social Issues of the Human Genome Project. He has taught philosophy at Georgetown and Johns Hopkins, and law at Georgetown and the University of Wisconsin. His current research focuses on ethical and legal issues in human stem cell research and end-of-life decision making, as well as on more theoretical problems in political philosophy and philosophy of mind.

GRETCHEN VOGEL is the Berlin correspondent for *Science*. Her areas of interest are science policy in Germany and Central Europe, developmental biology, and evolution.

RICK WEISS is a science and medical reporter for the *Washington Post*. He came to the *Post*'s health section in 1993 and moved to the national desk in January 1996, where he covers genetics, molecular biology, and other topics in the life sciences.

FRANK E. YOUNG has served in a variety of governmental positions, including as commissioner of the Food and Drug Administration (1984–1989), commissioner on the World Health Organization's Commission on Health and Environment (1990), and assistant surgeon general (rear admiral, upper half) of the Department of Health and Human Services (1984). He is the former dean of the School of Medicine and Dentistry at the University of Rochester and is an ordained minister in the Evangelical Church Alliance.